Table of Contents

Introduction

Working with Scrum can skyrocket your project's success to new heights. Many project managers wonder if it's really possible to do more with less. You might have the same questions racing through your mind. Is it really possible to finish projects with fewer resources? Is it really possible to make projects more manageable, more efficient, and more fun, with a single project methodology? The answer: yes, yes and … *yes*. In the various projects I embarked on as either a project manager or team member, one thing became evident: the methodology employed by an organization has an immediate impact on the team's efforts and results.

As a project manager or aspiring project manager, your success is your team's success; and vice versa. I have never seen a professional who didn't want to be part of a high-performing team and share in its wins. Professionals who were ever part of a highly-efficient and top-performing team can relate to this. In such teams, the collective energy is immense, leading to constant improvements in the products in development. Amazing results don't just "fall from the sky;" it's all cause and effect. And if cause and effect aren't adequately managed and channeled, they can delay or even block success

Without a doubt, a team that excels in cooperation, a team that flows, is one that achieves tremendous results with less effort. But a team on its own is not enough in the continuously developing world we live in. A team without the tools, techniques, and the framework to adequately deal with this dynamism, is like a ship without a rudder.

Fortunately, many visionaries, project managers, and entrepreneurs have woken up to the fact that the traditional way of working isn't suited for today's needs. Thus, they have developed new innovative ways to deal with projects from start to finish, called *agile methodologies*. Many leading companies, such as Microsoft, Apple, and Amazon, use an agile approach to tackle their projects adequately. These big tech companies are aware that continuous technological developments make such a flexible approach to handling projects necessary—maybe almost vital—to surviving.

There are various agile methodologies present today, such as Scrum, Kanban, and XP. Scrum is the most popular method or framework, and this book will make clear how you can implement it in your organization or company for ample project success.

With Scrum, you are given a framework with which you can develop various products that are part of a project. The Scrum framework and methodology started out being used for IT-projects exclusively. Today, things are different. Scrum is used for every kind of project: from transportation to agriculture to engineering projects. Now, isn't that amazing? Thus, I'm adamant that learning about Scrum will be of much use to you and your team, no matter what field you are in.

The Scrum approach makes competitors using traditional methods look like snails, struggling day in, day out to move the project forward more swiftly. Taking on projects iteratively optimizes predictability of results, mitigates risks—or sometimes even eliminates them—and makes you and your team more efficient. This all falls in line with the three main pillars Scrum is based on, namely transparency, inspection, and adjustment/adaptation.

In this book, we delve deep into the most up-to-date manner of implementing Scrum. That may sound overwhelming, but don't worry. I'll simplify matters as much as possible and cover the essential aspects and processes of Scrum in an easy-to-understand fashion, even for complete beginners. And don't worry, this book also includes some more advanced aspects of Scrum for you more seasoned project managers.

With my experience managing different teams in several industries, I have faced many failures and setbacks in my work as a project manager. Truth be told, I'm not a "born" manager, not at all. But with hard work, consistency, and determination, I distinguished myself from most project managers. Doing so was impossible without the use of Scrum.

I want to help others do the same. Therefore, I aimed to write this book without the fluff, but with accurate and practical information. Information you can apply from the get-go to help you move forward as a project manager. Besides, the book will include various examples, expert advice, and real case studies to give a more precise image of the reality around Scrum.

In the first section of this book, we start by giving an in-depth explanation about how you can get started with Scrum. We explain what it is and why you need it. The second section outlines the Scrum process from start to finish. You'll learn about Scrum teams, breaking down a Scrum project, Scrum Artifacts, and much more! In the third section, I hand you the necessary Scrum tools, tips, and other essentials to skyrocket your projects' success. This is done by looking at Scrum metrics, how to excel in a specific Scrum role, common mistakes, and software tools you can use.

So, what are you waiting for? Read forth and set yourself apart from the crowd!

Chapter 1: Project Management: Past and Present

Every innovation has its history. The same goes for methodologies to manage projects. Nowadays, more organizations adopt what we call *agile methods* for project management. When you ask someone to describe the term *agile*, you're likely to get multiple different answers. Therefore, it is helpful to take a look at the origins of this agile-way of working. That's precisely what we'll do in this chapter. Furthermore, you will learn more about various agile concepts, its components, multiple benefits, and much more.

The Origins of Agile Methodologies

Before organizations practiced agile methodologies, they employed the so-called "waterfall method" to get through projects. Winston Royce first mentioned this waterfall method in his paper, *Managing the Development of Large Software Systems*, that he published in 1970. Royce proffered a diagram for software development, similar to the one you see below:

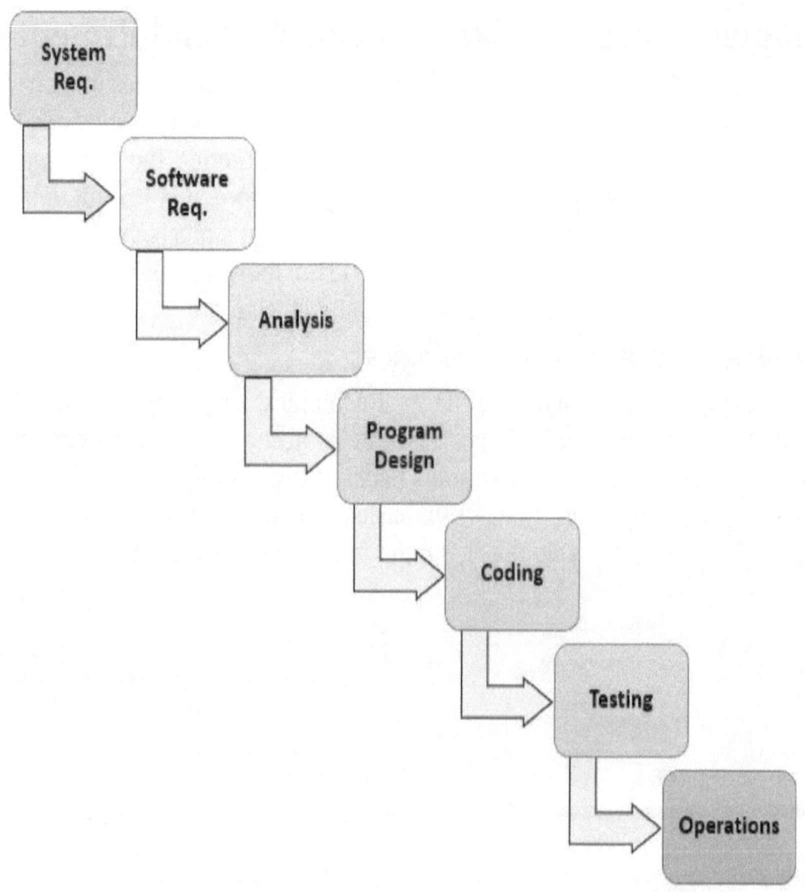

In the chart, you see the various phases of software development. Starting with the System Requirements and Software Requirements, it was possible to come to an analysis. Afterward, it was possible to design the program based on this analysis. When the design was ready, it was time to start coding and testing the software. Finally, the software would go to operations and would be used. As you see, each phase flows into a new one, without going back to a previous phase, similar to a waterfall. What I found particularly interesting to note is that Royce himself didn't see this method as optimal. This became evident in his paper when he described a more iterative and step-by-step approach of going through these phases (i.e., agile).

There is a reason the method goes through the phases like a waterfall. Remember that in the 1950s and 1960s, computers were as big as a house! They had various intricate parts that were hard to work with and needed professionals to change any of these parts. These computers were very time-consuming and complex to develop. Thus, the waterfall method was introduced to tackle this endeavor. There are numerous challenges involved by practicing this method. In the waterfall method, there are specialized people per phase. Just think of business analysts, architects, and designers, developers, QA specialists, and infrastructure specialists who did the deployment. The problem is the gaps between these activities and the need to transfer information between these phases. A lot of documentation was developed through each stage. The business analyst starts with the documentation and passes it to the designer who adds to it. Analysis and design together are what we call "Big Design Up Front" or, in short, BDUF. When the design is ready, the document is passed on to the developers. When they are done with the product, they hand it over to the testers. Afterward, they hand it over to the IT-staff who help the customers implement the software.

What usually happened was that when the developers were busying themselves with the program as described in the documentation, they found that what they had to create didn't work out. Then a change needed to be made. And this change was not found in the documentation made earlier. Now we have code that doesn't match the documentation—the same documentation that took the analysts and developers a lot of time to craft. Therefore, there had to be a way to go back to this documentation and improve upon it, remove parts, or add new elements.

Furthermore, within the waterfall method, *milestones* are generally used. These are detailed with a description and the date by which they should be achieved. However, in practice, it's challenging to always stay within the specified time frame. Sometimes more time is needed, making every defined milestone inaccurate. Also, the professionals in these phases didn't talk with the other professionals that weren't in direct relation to them. For instance, there is no clear communication between the testers and designers or analysts and the IT-staff.

It should be clear that this method is far from ideal in the rapidly changing digital era we live in today. Fortunately, there's a new, more agile approach to deal with projects. The new procedure was mainly created in response to the developments of computers in the 1980s and 1990s. Computers became easier to employ, and Software as a Service (SaaS) and the internet became available. In 1986, Hirotaka Takeuchi and Ikujiro Nonaka wrote a paper in the Harvard Business Review named *The New New Product Development Game*. In this paper, they delved into the phases of the waterfall method, namely: analysis, design, develop, test, and finally deploy. Besides giving more detail about the steps, they pointed out something very innovative. They showed that these phases don't only work on their own. In fact, the stages they proffered showed that the phases should have a very distinct overlap. Thus, information was better communicated between the professionals in each stage. For example, the analyst actively checks on the design and even looks at the development, from time to time. The paper suggests that the analysts should be around more frequently during the development phase, so that even at the end of the development phase, there is still a business representative (analyst) who can deliver the appropriate information.

Takeuchi and Nonaka described this process as being like a "scrum." This was derived from the Rugby term, when all players are linked together. Being linked together, they all work toward a common goal: trying to push the ball down the field. This symbolizes a team walking through the phases together, all the way to the end. Furthermore, from this paper, many processes, methods, and frameworks were developed and enhanced, such as Scrum by Jeff Sutherland and Ken Schwaber, Extreme Programming (XP) by Kent Beck, Kanban by Taiichi Ohno, and many more.

Fast forward to the 21st century, and current "thought leaders" have concluded that they're striving for similar goals. They are trying to achieve the same objectives to make processes more efficient, effective, and worthwhile. They found out that a lot of what they were doing in their methods has, in fact, a connection with other methods. The methods were more interconnected than the thought leaders imagined. Thus, because many of the core values in these methodologies were the same, they came together to bring to life what's called *The Agile Manifesto*, published in 2001. This manifesto introduced the world of project management to values it was in dire need of. It outlines that individuals and interactions are more important than processes and tools. But processes and tools still have a place; they just shouldn't intervene with, for instance, communication with stakeholders or professionals in a project. Besides, the manifesto values working software more than having comprehensive documentation. Don't get me wrong, documentation still has its place, but the actual, pure-practice of working with the software has the upper-hand. Customer collaboration is more significant than contract negotiation. Sometimes things change. Therefore, we must be able to have some form of dynamism in contracts; this isn't possible without adequate customer collaboration. In an agile approach, we value responding to change more than following a strict plan. No doubt, the adage, "If you fail to plan, you're planning to fail," sounds clear in our minds, but planning shouldn't be too rigid. Rigidness is what prevents organizations from moving through projects quickly. Moving quickly isn't possible without the flexibility to respond to changes swiftly.

Sometimes finding a more balanced approach between the values is necessary. In various organizations, people use tools such as Zoom or Skype for Business to make communication possible with remote workers or external stakeholders, using technology to support a focus on individuals and interactions. Therefore, extremes should be avoided. Each project should be evaluated individually to figure out how much each value should weigh.

The Key Agile Concepts

To develop a correct *agile mindset*, you should understand various agile concepts. The first concept you should know about within the agile-approach is **short feedback loops**. In the waterfall method, a customer may see no product for months, maybe *years* on end. If it turns out that the customer wanted a different feature or perhaps an entirely different product, it's too late. A key concept within agile is to get information to the customer as soon as possible, so they can find out ways to move forward in the shortest possible time. Thus, it's possible to fix things while the project is still running. The same goes for professionals, such as developers and testers, or even business analysts. With an agile mindset in place, every team member thinks "big" about the end product, but they work "small," so they can fall and get back up rapidly and learn from these missteps swiftly every single time.

The second concept is the so-called **just-in-time** way of gathering requirements and finishing the design. Frequently, developing software is compared to building a house. Honestly, it's nothing like building a house. There is no need to have complete blueprints outlining every intricate aspect of design from A to Z. What I've seen in practice is that development works exponentially better when working from a kind of "to-do list," as simple as that may sound.

The third crucial concept in agile is **delivering incremental value**. We cannot make a product entirely from the get-go. Thus, we make various parts of value along the way, working toward the end-product. Afterward, these incremental products can be discussed with customers or other stakeholders. Their feedback can be gathered, making room for improving this part of the end-product. The goal is to have incremental products that are ready to be released to the customer. This means, for software development, that it's integrated and documented thoroughly, the programmers have finished the code, and the program is tested and deployed.

The fourth concept is the **maintainable work rate**. You've probably experienced projects where things start easy, you work a little bit, and then find that a deadline is coming up soon, which makes employees work 80+ hour weeks all of a sudden. With agile, there is more control regarding the effort of employees. This effort should be equal during the entire project as much as physically possible. This way, employees don't get burned out, and they deliver better, more predictable results. Bottom line: you have to keep a pace that employees can maintain to get the best results for your projects.

Within agile, there is a concept of **Lean hierarchy and self-organizing teams**, and this is the fifth concept I'll address. This means that a few people make decisions instead of having an extremely slow hierarchy, where too many people have to say something before a decision is made. Doing it the *Lean* way means that it takes significantly less time when decisions roll out.

Regarding decisions for specific projects, self-organizing teams are essential. Therein is the central idea that the team is in the best position to make concrete decisions for continuing the project. Without self-organizing teams, there's usually a manager who makes these decisions while they aren't even aware of the intricate, crucial details within a project. Empowering self-organizing teams and allowing them to make decisions helps garner better results.

Values like **respect, collaboration, trust, courage, and transparency** are essential within agile methodologies. These give you the "agile mindset" to support the projects at work and are critical for **continuous delivery**, which is the sixth concept. When managers trust their self-organizing teams and there is a frequent enough dialogue, this makes projects very transparent. Delivering in short feedback loops makes continuous delivery a crucial concept. Any time the team builds something, it can take it from building and development to deployment as fast as possible. This comes in the form of continuous integration. For instance, as soon as I write some code, this is shared with other developers through a server.

We live in an age where most projects still "manage" change. In agile, we instead **embrace change**, then try to manage it. This is the seventh concept. Especially for customers, we always need to be willing to change things around in the product if it isn't satisfactory. Thus, changing becomes an integral part of the entire project, making it better to embrace change than manage it. Inspecting and adapting is necessary for any agile approach, be it Scrum, Kanban, or XP. This happens in the tools, but also the incremental products you deliver and even the team itself. These approaches are frameworks that demand your input. You have to fill them with what makes sense for you. Otherwise, embracing change will be a long, poor, tiresome road.

Scrum: How It All Started

Scrum was founded by Jeff Sutherland and Ken Schwaber approximately 25 years ago. After his military career, Jeff studied medicine at the University of Colorado. At university, he developed an interest in something utterly different than medicine: namely, IT. Schwaber, on the other hand, started his career early as a software engineer. Both gentlemen had a vision for a quicker, more reliable, and more effective method for creating software. They were frustrated with the inefficiencies present in the waterfall method and worked effortlessly to find a new way to tackle projects.

Sutherland, in his book *Scrum: The Art of Doing Twice the Work in Half the Time*, gives examples of these inefficiencies. For instance, take the digitalization project conducted by the FBI in 2006 for a new program called Sentinel. The program aimed to get rid of the paper processes to make room for digital operations. Any idea how much the project would cost? An immense $451 million. According to the contractors, the entire program, systems, and underlying processes would be up and running somewhere in 2009. Fast forward a few years later, from 2006 to 2010, and there was no working program, and the considerable sum of $451 million was already spent. In the following months, the stipulated project cost was far-exceeded, and expenses from the contractor didn't seem to stop. But there was no other option, because the project was already halfway, and it had to continue. Well, at least that's what they thought. Thinking this way is faulty. It's built on the premise that we should continue with a project no matter what happens because we've already invested a significant amount of time, money, and energy. This is called *the sunk cost fallacy.*

Long story short, the team hired by the FBI kept at it and estimated that the project should only take about six to eight more years. Oh yeah, and did I mention that an additional $350 million of taxpayers' money was needed? Although genius engineers, managers, and analysts were relentlessly working on the project, things didn't work out. Their process involved the same phases mentioned in the waterfall method. So, they would gather requirements, analyze these requirements, make a plan for the contractor, and add details with various features needed for the program. A group of very talented professionals from various fields worked for days, weeks, and months without stopping. After the plan was ready, they spent more days, weeks, and months planning the process for implementing the plan. To make this clear, they made sure all documentation was easily accessible, was nicely designed, and included various shining graphs and diagrams detailing the tasks and amount of work necessary.

The graphs and diagrams indicated what part of the plan should be implemented first before moving on to the next part. They included various stepping stones to get to the destination. After a couple of stepping stones, a milestone was highlighted. Between the tasks and stepping stones, the deliverables were made apparent. With the introduction of numerous software tools for creating graphs and diagrams, it was easy to keep adding more elements to them and making them more complex along the way. Eventually, you can't see the forest for the trees. Besides, these graphs and diagrams may look fancy, but they're nearly always wrong. Why? Because they aren't suited to a project environment that isn't dynamic. Developing software is a continuous and changing process. Doing so in a vacuum doesn't end up well, not at all, as we read from this example.

The managers for this project at the FBI thought they had all the resources to make the project a wild success. From great talent, as mentioned earlier, to advanced technology and software systems. But something was missing. Something that may seem subtle, but it's a subtlety that makes all the difference: people worked and planned wrongly, i.e., their methodology was invalid. Such a great and intensive project was never feasible with the old, traditional, inefficient way of working. There was a new, innovative, efficient way needed.

Fortunately, after a lot of blood, sweat, and tears, there was a time of joy. Finally, one of the genius managers realized that the plan made a couple of months ago was now a work of fiction, due to continuous changes along the way. When the manager took a closer look at the raw development process, and its similar products and services, they knew the plan was not at all valid. Eventually, they discovered that a new way to manage projects was vital. Thus, they were introduced to Scrum, one of the only methodologies to make these data-driven, complex projects a wild success. They realized that the way of managing projects in the past isn't applicable anymore. Going forth with this waterfall approach will cost a tremendous amount of resources.

Furthermore, the ideal outcomes are often not achieved. In the FBI example, the waterfall method of doing things cost hundreds of millions of dollars and a lot of resources. In the new, agile way of working, people can do more in less time. People can learn from mistakes and readjust in less time. And people can achieve better results in less time. Doing more with less is the motto within agile methodologies.

It may sound like a fantasy, but various organizations show amazing results with this approach. The agile-way works, and it works for all kinds of organizations. Sutherland and Schwaber found the method by looking at *how* people do their work, instead of listening to what they *say* they do. Both gentlemen studied ways to make projects more durable by looking at studies conducted around project management. Also, they took a closer look at how other organizations manage their projects. Doing so showed them a pattern of what works during projects and what doesn't. They knew there was something behind the successes of various businesses around the world.

They concluded that most successful organizations managed their projects in a more iterative manner, such as Amazon. Now, Amazon is well-known to implement Scrum not only in smaller projects but even in various business layers with larger projects. Do you want to learn how Scrum changes organizations for the better? Do you want to help your organization move forward quickly? Do you want excellent project success? Adopting a Scrum methodology will pave the way for excellent project outcomes. Let's have a more detailed look at what Scrum is about.

Chapter 2: Scrum: What It Is and Why You Need It

Now that we have a clear understanding of agile, its origins, concepts, values, and benefits, I hope it is evident why you need a more agile approach for project success. In this chapter, we'll delve deeper into the agile methodology —or framework—of Scrum. The difference between Scrum and other agile methods is that it's the easiest and most flexible method to implement. Contrary to what many project managers think, Scrum is not just for software development projects. As you read earlier, Scrum originated from the software development world, but it is nowadays adopted in nearly every reasonable-sized organization, regardless of the industry.

The Basics

Scrum doesn't require any advanced mathematics or rocket science. You could jot down the most basic Scrum elements on a Post-it Note. Take a look at the essential elements on which Scrum is based:

- **Plan the short term in detail, without forgetting the long term.** Within Scrum, we know that detailed planning and schedules can only be used effectively in the short term. This is quite the opposite of traditional methodologies, where detailed scheduling takes place for events that are "light-years" in the future. This doesn't mean that a successful Scrum team doesn't think long term. Not at all. Team members think long term, but know that day-to-day scheduling of tasks in the short term will help them get closer to the long term objectives. Scheduling the long term in detail is far from possible because too many variables can change over time. It seems so logical, but many (project) managers seem to think the opposite.

- **Self-organizing and multi-disciplinary teams for the win.** If you want to win big in business, you need a great team. Especially within Scrum, the team plays a central role and works more innovatively. It is a team that decides what it will work on, how long it will take, and when it is finished. No manager has any power to enforce their wishes on the team. The team consists of all the professional disciplines needed to get the job done. Team members know the work, and the manager usually doesn't. So, wouldn't it make more sense for these professionals to figure out what to work on and when things should be finished? A rhetorical question, of course. Team members within the Scrum framework craft the plan, organize the tasks, and carry out the work.

• **Chunking projects into reasonable sizes.** Whenever you are faced with a large project, it's difficult to swallow the "elephant" in one bite. Instead, make reasonable-sized chunks to get it down the throat. "Chunking" or splitting significant tasks into smaller parts is crucial during any Scrum project. Within Scrum, projects are divided into short sprints, where you develop and deliver valuable incremental products for the customer.

• **Transparency is key.** This is what Scrum is all about. With Scrum, no team member can "fool" another team member in terms of conducted work. This forces collaboration and helps the team members get the job done well and in a timely fashion. If you want to get things done and keep good relationships, there is only one way: communicate transparently. Only by communicating transparently can you achieve what you want in the best, fastest, and most enjoyable fashion.

• **The necessary feedback-loop.** When you offer your customers the opportunity to give feedback, you know exactly what is going on with them and you can respond. In this way, customer feedback plays an essential role within your company. When customers and stakeholders regularly provide feedback on the product increment, this will hopefully bring about a better result.

• **Don't forget to communicate!** The team regularly discusses with each other whether it can improve the working method, making it more and more effective. The intensive cooperation, self-organization, feedback, and fast results almost inevitably lead to more pleasure when doing the work.

As said, there are no advanced mathematics anywhere to be found. But still, millions of project managers manage ineffectively. People who hear about Scrum for the first time often say that they are already using all these essential elements. However, being able to "practice what you preach" appears to be the obstacle. The effectiveness of Scrum stands or falls with implementation. Experienced Scrum professionals know that it's not about whether you employ Scrum, but how well you use it. The outcome is the mark of the success or failure of the team's efforts, scheduling, and planning.

Scrum is still finding its place in different industries besides IT. Some industries have taken up Scrum very well, while others remain behind. In my experience, I have come across things that work slightly differently than in the traditional software industry. These points of attention are indispensable for the broad application and implementation of Scrum. To figure out if your organization is ready to use Scrum, answer the following questions, and count your points!

- **Are projects always done on time?**
    - Yes. (1 point)
    - Once in a while. (2 points)
    - Seldom. (3 points)
- **Do the projects never get mixed up?**
    - Yeah, that's right, I can always focus on a single project. (1 point)
    - Nope, I have to divide my time and focus on various projects. However, I still have an overview. (2 points)
    - No, it is all mixed up, and the work feels too fragmented. It almost seems like the organization is some "project carousel." (3 points)
- **Does your organization usually think and work in projects?**
    - Yes. (1 point)
    - We are working on a more project-based approach. (2 points)
    - We want to work on a project basis but are not yet geared to this. (3 points)

- **Would you say that other team members are motivated?**
    - Yes. (1 point)
    - Once in a while. (2 points)
    - Rarely. (3 points)
- **Do the team members always deliver what the customers or stakeholders demand from them?**
    - Yes. (1 point)
    - Things often have to be redone; this costs a lot of extra energy. (2 points)

o Stakeholders/customers are not always satisfied with what we deliver. (3 points)

- **Would you say that other team members exchange a lot of knowledge and skills?**

  o Yes, we learn a lot from each other and use each other's knowledge and skills. (1 point)

  o Once in a while. (2 points)

  o No, way too little. (3 points)

- **Does the team usually do the most important things first, and do the team members not allow each other to be distracted by side issues?**

  o Yes, we always tackle the most critical issues first. (1 point)

  o Once in a while. (2 points)

  o Many colleagues are busy with things that make me wonder if they are critical. (3 points)

- **Do you think that stakeholders are actively involved in the project?**

  o Yes, during the project, there are several moments when stakeholders give feedback and indicate their desires and wishes. (1 point)

  o Sometimes, because we analyze the stakeholders and invite them once in a while to check the progress. (2 points)

  o No, because I think that too many times, stakeholders are hardly involved in the process. (3 points)

- **Would you say that the team is flexible and can quickly adapt to changes in the desires and wishes of the customer or the project environment?**

  o Yes, if anything would change, we can make adjustments in the short term without causing (m)any problems. (1 point)

  o We can make adjustments, but that often requires a lot of art and work on-the-fly. Sometimes we even have to redo advanced projects. (2 points)

- We work according to a schedule that is difficult to adjust en route. (3 points)

- **Is your current working method providing much pleasure for the entire team?**
  - Yes. (1 point)
  - Once in a while. (2 points)
  - Rarely. (3 points)

After you have answered the questions add up your score! The first answer option is worth one point, the second two points, and the third is—you guessed it … three points. Now, let's see where your organization is currently at:

- If your score is 10-11: Your organization is already great at applying Scrum elements. Maybe the organization has several years of experience practicing Scrum or a different agile methodology. These skills can be sharpened by using more advanced techniques, described in a later chapter.

- If your score is 12-21: You're doing fine, but there is a lot of room for improvement. The organization is most likely implementing various Scrum techniques and elements, but they don't go too well yet. This guidebook will help you improve on the probable pitfalls.

- If your score is 21-30: Your organization needs to change. A considerable change needs to happen to save your organization's project because working in the same ineffective and inefficient way will be detrimental to the organization and the people involved. This book will set you up to implement Scrum to a very good degree. Thus, your organization can change things around for the better.

What Is Scrum?

Scrum is an agile methodology to tackle projects. It is based on a fundamentally different vision of working together so that many traditional project pitfalls are avoided. Most people have experienced or heard about large project teams that, after months of work, deliver half-finished products that no one is waiting for. With Scrum, we do the exact opposite. We cut the large project into pieces and finish small parts every few weeks. This is done in sprints: relatively short periods of two-to-four weeks, during which we realize and deliver parts of the project. Customers can see a more rapid and better result, and can give immediate feedback. This allows us to respond much better to customer requirements.

Scrum is more than just a vision. It's a practical method to work productively with a dedicated team. The core of Scrum is explicit and consists of roles, ceremonies, and lists. There are clear roles so that everyone knows where she/he stands, fixed ceremonies in which the team comes together, and a few handy lists that replace extensive, highly-inefficient plans. It is essential to have a good understanding of this and use the Scrum elements in the right way: only if you do so correctly is it possible to say you're working with Scrum.

Scrum consists of three roles, four ceremonies, and four lists. Scrum has three distinctive roles. You form a Scrum team with people who together can do the most substantial portion of the tasks at hand. The group consists of an average of seven people, plus or minus two, often from different disciplines. The team is self-organizing. This means that the team members decide together how they want to carry out the tasks and divide the work. A Scrum team has *no* project manager. You might think, "No project manager? Then how do the team members know what needs to be completed and when?"

Well, first of all, the Scrum team has people from all three roles, namely the Scrum Master role, Product Owner role, and Development Team role. The latter is different than the phrase "the Scrum team" because it doesn't include the Scrum Master and Product Owner roles. Instead, it contains professionals from various disciplines, such as business analysts, designers, and programmers, to take care of the tasks. Furthermore, the Product Owner is the delegated principal for the project, i.e., the one who gives the project to the Development Team and has close contact with the customer(s). She or he makes an inventory of the wishes of the internal or external customer(s) and translates this into a clear assignment for the team. The Product Owner monitors the job, the priorities, and the preconditions and makes decisions where necessary. The Product Owner "owns" the product or the content. And then there is the Scrum Master, the facilitator of the Scrum team. The Scrum Master guides the team so that the process runs smoothly. The Scrum Master is therefore responsible for the quality of the process: ensuring that the

Development Team takes the right steps and that ceremonies take place in the right way and at the right time.

Scrum ceremonies occur as four different types of team meetings. The lead time of the project is divided into equally significant periods called sprints. Every new sprint starts with a sprint planning meeting, in which the team determines how it can achieve the most important goals for this sprint. During the sprint, stand-ups (i.e., daily "Scrums" or stand-up meetings) are regularly held. These are short interim discussions of no more than fifteen minutes, during which the team members tell each other the progress of the tasks. Standing keeps the team members going and prevents people from leaning back and losing interest. At the end of every sprint, the team presents everything that was made in this sprint to the Product Owner. This is called the sprint review. Sometimes other stakeholders are also invited, such as colleagues, customers, or directors. The fourth type of ceremony is a final retrospective meeting. In this meeting, you and the team look back on the process, so that you can improve team performance in the next sprint.

When I started with Scrum in my previous job, we did ask a Scrum coach to guide the teams during the first sprints. The basis of Scrum is simple, but applying it well in practice is a craft where coaching is more than useful. The Scrum coach then immediately trained several enthusiastic colleagues and me to become a Scrum Master.

Now, we arrive at the last indispensable part of Scrum, the lists. The four Scrum lists are nothing more than visual aids. In Scrum, you usually display the lists on flip charts with Post-it Notes, that show what the team is working on. The first list is the *Product Backlog*, the overview with all requirements and wishes for the entire project. With Scrum, you no longer have to write an extensive long-term plan, but the Product Owner makes an inventory of the components that must be worked out for this project. For each part, a separate Post-it is placed on the Product Backlog, and these are called the "backlog items."

At the start of every sprint, the Development Team selects, along with the Product Owner, the items from the Product Backlog that the team will realize in the coming sprint. These Post-its move to the second list: the Sprint Backlog. On the third list, a "definition of done" is then written for each item. These are the requirements that a thing must fulfill to be considered "done." The "definition of done" answers the question: What exactly will be finished and achieved at the end of the sprint, and what does that look like?

Formulating this with the team creates a shared image of what you will deliver at the end of this sprint. Also, there's an extension to this list that's usually ignored: "the definition of fun." This is a list of conditions for making and keeping work within the Scrum process fun. An essential element of preparation for the team is the Scrum board, which the items of the Sprint Backlog are placed on. The purest form consists of the columns "to do," "busy," and "done." The Scrum board can be digital but is usually made on a physical whiteboard or flip chart.

Is Scrum Needed?

There are many reasons why Scrum is needed when you work on various projects. Below I'll give multiple reasons why you need to use Scrum for your projects as soon as possible. However, please take notice: I could name many more reasons to use Scrum for projects. And many other reasons are scattered throughout this book. For this chapter, the reasons below will suffice.

The first reason is that you gain more value from your resources, such as time and money, but also talent. With Scrum, it is always clear what team members are working on, and product increments are delivered as fast as possible. The most important parts of the end-product are usable early on because of this. Therefore the "Time-to-Market" of a product can be drastically shortened.

The second reason is that Scrum gives the team more control over the whole product creation process: from beginning to end. Scrum is an empirical process, and by forcing yourself to get feedback as quickly as possible, a lot of information is garnered. This information also gets better and better, and stakeholders can employ this information to help move the project forward. This contrasts with traditional methods, in which setbacks usually appear when the project is "nearly finished." Most of the time, it's then too late.

Furthermore, the third reason is delivering higher quality products for customers and/or other stakeholders. By asking customers and other stakeholders for their opinion during every sprint review, you never lose sight of the users' wishes and desires. This gives you an edge over other organizations that don't have this continuous process of checking and validating projects in place. The Scrum approach helps to get a better understanding of what is really bothering customers. In the process, everyone learns what's important. By doing production-based product increment deliveries every time, the attention to detail and quality is great. This is far superior to making a ton of assumptions in the beginning, and then facing the problems further down the line.

Another good reason to use Scrum is that it allows you to explore uncertain, more complex projects without losing too many resources. Instead of having long and expensive documentation done by several external consultants, it can be gratifying to put a Scrum team to work on several sprints. After a few weeks, you'll know whether a new product is feasible. If not, better luck next time, at least you learned something. If it works out, which happens often, then you're immediately ahead!

Besides, Scrum results in less bureaucracy. In line with the previous point, many organizations have become much more cautious about spending resources like money and time due to bad experiences in the past. Thus, procedures have been developed to avoid these. After a while, these procedures have taken on a life of their own. Therefore, frequently, these procedures take a long time. Mainly because work is waiting for approval from Change Advisory Boards and the like. With Scrum, these obstacles disappear.

As you can see, the basics and reasons for applying Scrum are straightforward. No advanced math to be seen. However, appearances are deceptive. Behind the simple roles, ceremonies, and lists, is a fundamentally different way of working. The combination of these factors makes it work. Always remember that a too dogmatic approach to implementing the concepts is far from ideal; for yourself, for your team, but also for the organization. Therefore, you must not blindly apply all of the Scrum elements we discuss in this book, without realizing the value for your specific project. Instead, you should evaluate your project and your team's needs and cater your application of Scrum to those needs. Besides the essential Scrum elements described earlier, there are many more elements you can use during a Scrum project, as you'll see in the coming chapters. Add to the basics by selecting other concepts and elements that you're pretty sure will get your team moving the most. Don't be afraid to add a new perspective to elements where you think and feel it's necessary. Without further ado, let's delve deeper into Scrum!

Chapter 3: Scrum Roles and Responsibilities

By employing the waterfall method, it isn't until the last phase that your customers get to interact with the product. Then is the time you get to know if what you produced is what they were looking for. Because it is right at the end of the project, what can be done if the customer isn't satisfied? If a few requirements are outdated? Or, if a couple of elements are missing? This is an absolute recipe for disaster because people don't know what they want, until they interact with it!

Scrum is great for getting the ball rolling quickly in projects. Scrum functions as a framework for self-organizing teams to conduct projects effectively and efficiently. It consists of three categories you must know about, namely: roles, artifacts, and events. During the explanation of these categories, I will add additional insights for properly implementing them. These insights may or may not be based on The Scrum Guide by Jeff Sutherland and Ken Schwaber, but they seem essential to me when dealing with Scrum processes. This chapter is dedicated to the roles in Scrum.

Product Owner

In Scrum, there are a couple of roles. The first role is the Product Owner. The Product Owner strives to maximize product value; that is their responsibility. Whatever we do in a project should create, deliver, and keep value for the customers and organization. The Product Owner makes sure this is the case.

Furthermore, she/he manages the Product Backlog, which is the only document or source where all the requirements are listed. It is the Product Owner's job to make sure the Product Backlog is well-formed, makes sense, and is prioritized. Also, the Product Owner represents the customers by often communicating with them. Finally, they make so-called go/no-go decisions, for what will be released and what will not. Finding the right person to fulfill this role is difficult, because of the wide variety of skills necessary to complete these tasks correctly.

The Product Owner takes ownership of the product and is responsible for its successful realization: on time, within the stipulated budget, and with satisfied customers. All three have to be met for a successful project. Missing one of these three aspects means missing delivering a great product.

Scrum preaches simplicity and transparency, and the role of the Product Owner is an excellent example of this. After all, there is only one Product Owner, and that is the same person during the entire project. This way, everyone knows who makes the decisions. Decisions about the direction of the product, that is.

Moreover, every Product Owner only has one product under her/his care, and that is all she/he focuses on. This gives a clear image to the Product Owner and results in a more significant commitment. Why? Because the Product Owner will spend all their time on directing the product until it is finished. Typically, the Product Owner will spend half of the time with the stakeholders and the other half with the team. The role of the Product Owner is a full-time job.

Although various skeptics might disagree, the Product Owner needs to be full-time, because the work that needs to be done is extensive. The Product Owner is not only concerned with the team, and perhaps more things are being developed than just software. The Product Owner is involved with stakeholders for a considerable part of the time. This includes all kinds of activities, such as discussing with customers; coordinating with the marketing department; making sketches to get a better picture of the target audience; and coordinating the budget with the management.

It is important to note that the Product Owner shouldn't cause a bottleneck. This will result in unnecessary costs because the professionals have to wait to get to work. So, the Product Owner must have enough time to be present, to show the way, to motivate people, and to repeat the vision.

It is the responsibility of the Product Owner to represent anyone with an interest in the product and to weigh the interests between all these people or parties, and to constantly decide what is essential. Indeed, I say "constantly" because, as everyone knows, the demands and wishes of these stakeholders are continually changing.

The Product Owner is also responsible for discussing with the Development Team about the implementation of the requirements and wishes of the stakeholders. The Development Team usually has several ways to interpret and implement a requirement. The costs can, therefore, vary considerably, and the Product Owner must understand that decisions are made by the Development Team as well. In the discussions with the Development Team, it is mainly about the "how" and the "what" in terms of cost.

With these two aspects of requirements: "what" and "how," the Product Owner prioritizes the needs or items listed on the Product Backlog. She/he weighs up the order in which the objects are realized. This is often done by presenting the estimates of the Development Team to the stakeholders, so that "quick wins" are put forward. Besides, it is possible that some requirements may even be canceled due to costs, for instance. When you are the Product Owner, prioritize first, then ask for costs. Cost estimates take a relatively long time, so pay more attention to the most vital issues. It is your primary responsibility to get value for money. You want to get a return on your investment (ROI) of time, energy, and money.

There is always the possibility that things won't go as planned. Thus, you could also mess things up. Just take it on the chin, these things can happen. The development of new products remains an uncertain and complex undertaking. After all, that's why we use Scrum. Scrum is not the guaranteed pathway to success. However, it is a guaranteed way to uncover all challenges, opportunities, and possibilities, as quickly as possible. Many successful companies use Scrum to get started and test the business case, even though the chance of failure is significant. If it works out well, they know they have won time, money, and other resources. If things flop, because the team will not be ready on time or there aren't enough funds, for example, then they know when to stop much earlier than in traditional methodologies. Therefore, even failing with Scrum is better than with any other way! Because you lose the least amount of resources, get back up quickly, and start working on the next thing in line without hesitation.

As explained earlier, the main task of the Product Owner is managing the Product Backlog. As a Product Owner, you are always (re)prioritizing the items on the Product Backlog, so that the most valuable things are placed at the top. These are the items with the highest value and relatively lowest costs. The Product Owner must always prioritize all requirements and wishes on the Product Backlog based on business value. These items should get an estimate of the effort required by the Development Team. This is called "backlog refinement," also known as "backlog grooming." The Product Backlog is very dynamic, mainly because the Development Team often demonstrates valuable, production-ready software, and stakeholders gain new insights as time progresses. However, variables like budget, market needs, and the actual use of product increments that are developed each sprint, have an influence too.

The Product Owner needs to have a knack for making the Product Backlog as valuable and understandable as possible. The Product Backlog is primarily a communication tool and can benefit a myriad of people in the organization.

Furthermore, it can be the source of many questions and discussions, but there is nothing wrong with that. It is better to ask questions straight away, rather than letting these roam around in people's heads. This would never be possible if the Product Backlog is in someone's drawer collecting dust: that's not the way to go. Instead, hang it on the wall or designate a whiteboard for it, to make it visible to anyone who has pending questions.

Also, nobody understands a Product Backlog full of jargon. Usually, if that is the case, it will lead to numerous issues, like poor acceptance by users, poor communication, and far less value for time, energy, and money. There are already enough organizations with a far too complex infrastructure. So, don't make things even more complicated, or it will bring forth many disadvantages. Instead, only list recognizable and usable points for users on the list, written in the words of a user. Go to production as quickly as possible and as often as possible (yes, go live!). Nothing gives more insight into the correctness of your decisions than the use of the product or product increment by real people.

Scrum Master

The next role is that of Scrum Master, which can, to a certain degree, be compared to project managers from traditional methodologies. So, just imagine yourself in this role. A Scrum Master is like a shepherd of Scrum because they need to know the concepts and implement the right mechanisms. They make sure everyone adheres to the agile values. As a Scrum Master, you should serve the Development Team, but don't tell them how and when they should do their tasks specifically. Also, if the team has problems that get in the way of producing excellent results, you should remove any barriers by identifying the issues and removing the obstacles.

Finally, as a Scrum Master, you also focus on resolving conflicts so that the team is back on track as soon as possible. The main difference between project managers is that you, as a Scrum Master, are there to empower the team. You aren't there to be in command and tell every team member what they should do. You should leave them to do their work because they're specialized in that work. A team leader can take on the role of Scrum Master, but this is usually the case when a team is more mature. When the organization is new to working with Scrum, it's difficult for you as a team leader to juggle between being a Scrum Master and leading a team. You may get too busy educating people on the process so that there is no room for other activities. Therefore, it may be advised to have a specific Scrum Master per team. When the teams mature more and you gain more experience as a Scrum Master, you can be the Scrum Master for multiple teams. How come? Because with practice the teams will become more self-aware, self-resilient, and self-actualizing. Thus, you will have less work as a Scrum Master in terms of educating the teams and putting out fires, so to speak.

One of the most significant responsibilities of the Scrum Master is that she/he makes sure the team members adhere to the "rules" of Scrum. A Scrum Master ensures that the Scrum process runs optimally. Therein, we find the most significant difference with the Product Owner, who makes sure a great product is delivered. Without a doubt, all roles in Scrum work toward a common goal, but they have various responsibilities. A Scrum project has one Product Owner, one Scrum Master, and one Development Team, all of which ensure that the common goal is achieved. What is the common goal? Well, it's to realize the Product Owner's vision, i.e., realizing the product and doing so as efficiently and effectively as possible.

You, as a Scrum Master, should make sure that the rules of Scrum are followed. During the process of completing tasks, impediments or obstacles will naturally come forth. Then you, as a Scrum Master, should step up to remove those barriers. Besides, the Scrum Master must convince people that working according to Scrum leads to better results. Thus, you can connect with more stakeholders and team members, making the process of production more manageable. To make things clear, everything that you should do as a Scrum Master stems from these responsibilities:

- Removing impediments or obstacles the Development Team faces.
- Making sure everyone adheres to the rules of the "game." Guarding game rules.
- Get people on board with Scrum, organize support for Scrum.
- Create positive change in the organization utilizing Scrum properly.

You, as a Scrum Master, are similar to a referee and coach. You are a referee because it's your responsibility that every team member adheres to the Scrum rules. You are a coach because you facilitate the entire process for all team members. You remove obstacles and help move the team forward toward project success.

As a Scrum Master, it is, therefore, vital to organize everyone's support for Scrum. Make it clear that Scrum is not the goal but the means to achieving the team's goal. Scrum is a very suitable approach to realize a product that is complex.

The person you should first have a discussion with and get on board is the Product Owner, no questions asked. If you don't have her/him on board, the process will be like climbing a huge mountain. When the Product Owner is ready to go, both of you can start motivating the team to follow Scrum by explaining why it would lead to better outcomes. As a Scrum Master, ensure that you don't take over the responsibilities of the Product Owner. The Product Owner motivates with regard to the product. And the Scrum Master motivates team members concerning the process of creating the product *using* Scrum. You don't wait to start the Scrum project until "everything" is ready and perfect. Scrum emphasizes the process of continuous learning, especially learning by doing. If you don't take proper action as the Scrum Master, the project will stagnate.

There are phases every Scrum team goes through when starting. In the beginning phase, everyone is busying themselves learning about Scrum, and the corresponding rules. After the rules are understood, you can only really *know* them by pure practice, i.e., doing a project using Scrum. After the team is more mature, in the second phase, it can add more advanced tactics or strategies to perform the projects more efficiently and effectively. Finally, the last phase is mastery, where the team performs projects according to Scrum, almost effortlessly. The aim should be to get to the second phase as soon as possible. There is no other way than consistently clearing projects with Scrum. Achieving mastery at anything might take years and years of hard work. The same is the case for the last phase. Thus, it is no problem to stay in the second phase for a more extended period.

Whatever you do, know that Scrum is a powerful concept, used by thousands of organizations worldwide. Don't think too swiftly that you know it all and can make your own version of Scrum. It is imperative that you garner support from management and that they understand that the investment in Scrum is a long-term investment. Make it evident that you, as a Scrum Master, can't fix everything alone and may need help from other professionals. Besides, the management and other stakeholders should know that Scrum is also a tool for meaningful discussion, to gain commitment from everyone. Obtaining commitment is done best with education. Give a workshop regarding Scrum or a presentation where you outline the benefits and case studies. Besides, you can regularly talk about the process. This last point is, of course, possible in the sprint retrospective, a Scrum meeting I'll soon address, but you could also have a chat when you feel some stakeholders show some degree of resistance to Scrum or changes in general.

Making sure that everyone adheres to the rules is far from easy, especially if your team is new to working with Scrum. People have habits to take care of their work, and these are difficult to unlearn. Scrum encourages cooperation and transparency, everything is visible, and nobody in the team has ownership of parts. So, it often happens that team members have difficulty with their new position in the team, and it is your responsibility to bring that to their attention and make it discussable. To further illustrate this, let's say we have two testers in the Development Team, and they have the ingrained habit of making a test plan based on a functional design. They are used to the developer delivering the software at the end of a release, and they test the software for a couple of weeks based on their plan. However, in Scrum, things are different. Because in Scrum, we no longer make a functional design in advance, and a tester will have to contribute something meaningful before the software is ready. What does a tester do in the first days of the first iteration? And what about the business analyst? These are all questions the Scrum Master needs to tackle. This is a role with many challenges, in which you can use all your experience, but especially all your persuasiveness and communication skills. Fortunately, you are not alone; after all, you share a common goal with the Product Owner, Development Team, and many stakeholders.

Scrum makes it possible to readjust quickly when things don't work out. If the Product Owner missed some requirements on the Product Backlog, or stakeholders don't want to collaborate, then that will result in much more pain when using Scrum than other methods. As a Scrum Master, you may be inclined to ease the pain and even pick up some work the Development Team is struggling with. However, don't do this, but instead, show and ensure visibility of the problems and only remove obstacles that are blocking the Development Team from doing the work. Don't do the work itself, but pave the way for the work to be done!

The rules of Scrum help you as a Scrum Master to bring issues to the surface. The daily stand-up meeting, for example, and the sprint review at the end of each sprint, give you every opportunity to get people to have their say. This information is essential to you as a Scrum Master. It allows you to find improvements. For instance, when there are problems in the realization of a feature that jeopardizes the planning, bring them up, and ask the team to discuss alternatives with the Product Owner. When a team member mentions a specific problem for many days, make sure there are some small tasks on the Sprint Backlog to tackle these issues; we will cover this in more detail later.

I've mentioned something about "rules" numerous times, but what are these "rules?" The most important ones are listed below, but don't worry if some concepts are still unknown. These will be addressed later. The rules are:

- In Scrum, there is one Product Owner. The Product Owner is responsible for the success of the product and has the mandate to make decisions regarding the product.

- There is one Scrum Master. She/he ensures that the Scrum rules and principles are adhered to, so that everyone can optimally concentrate on their task to make the product or product increment.

- In Scrum, there is one Development Team that independently develops the product. They do the tasks to get it in production.

- Scrum is done in sprints. These are continuous iterations of two to four weeks. Shorter is better. Sprints start and end on set days.

- All requirements and wishes for the product are recorded on the Product Backlog, which is managed by the Product Owner. The Product Backlog is always prioritized by value for the business, and the feasibility and cost of items on the Product Backlog are estimated by the Development Team because they know and do the work.

- A sprint starts with a sprint planning part 1 meeting, in which it is determined what will be dealt with in each sprint. Generally, these are the items at the top of the Product Backlog.

- Subsequently, in sprint planning part 2, the team—in the presence of the Product Owner—determines how the work is done and how they will progress the work.

- Every sprint, a number of tasks are moved from the Product Backlog to the Sprint Backlog. By doing this, these more significant Product Backlog tasks are split into smaller tasks the team can tackle on a daily basis.

- Also, making progress transparent is necessary for the Development Team as well. This is done through a "burndown chart."

- The Product Owner and the Development Team have made agreements. For instance, it's clear what is meant by a finished Product Backlog item. This is written down in the definition of done. "Done" means: so good that it can be taken into production.

- Every day the team holds a Daily Scrum meeting or stand-up meeting for not more than fifteen minutes. During this meeting, the Development Team goes through the tasks of the Sprint Backlog together, utilizing several standard questions. This meeting is public, and stakeholders can listen. They can only observe and must not interject during this meeting.

- At the end of the sprint, there is a meeting named "sprint review." Therein, the Development Team shows the results to the Product Owner and internal/external stakeholders and then receives feedback. The team only delivers the results that are "done," i.e., that can be taken into production by users.

- After the sprint review is completed, the team holds a sprint retrospective to pause and discuss how things went during the sprint. What went well? What can be improved? Make sure agreements are made, to enhance the process of the coming sprints.

Development Team

The third role within Scrum, as you might have noticed, is the Development Team. Within Scrum, everyone is developing something, so everyone is a developer (not necessarily a software developer!). All the professionals are part of the Development Team, be they an HR professional, software engineer, or business analyst. When you are on the team, these professional labels fade away, making you more aware and focused on the team's goal and working together. These teams are cross-functional or multi-disciplinary. So, they consist of the professionals necessary to come up with the product increments. That is what makes them self-organizing teams; the team knows design, programming/developing, testing, and any more skill sets needed to carry out the project successfully.

They should decide what they will work on and how they will do it—without much, or any, external help. In a Development Team, collaboration is crucial. So, if there is a problem with testing or more effort is needed in that area, a business analyst in the product team can jump in and lend a hand. Being collaborative means that you may give up some of your responsibilities or take on some extra tasks to accomplish the objective. Finally, the team shouldn't be too large.

The Development Team does all the work to turn a Product Owner's vision into a working product. They do all the direct work to complete the product. No one works directly on the product before or after the team starts or is finished. This means that the team gathers the requirements, performs analysis, makes a design, takes care of the underlying architecture, implements the product (increment), does the testing, installation, and documentation. This is all done by a multidisciplinary team of around five-to-nine people. Could Scrum work with more people? Well, not in the same team, as that leads to too much overhead. Adding people to a team is already less effective at eight or nine because you'll lose focus and commitment. What about using Scrum with fewer than five people? This could be possible, but then the Scrum Master or the Product Owner usually has a double role. That is far from ideal. You can't use Scrum with one or two people. What would one person do during the stand-up meeting? Of course, it is possible for one or two people to incorporate various Scrum elements into their processes, but not Scrum in its entirety, because Scrum is a team sport!

For example, if you are developing a communication system with the programming language Python, you may need an assisting architect, two Python developers, a business analyst, and a tester. If the communication system requires a fresh design or some form of interaction, then you add a CSS expert to the team. Various projects could consist of a wide variety of professionals. Even though the Development Team in Scrum is multidisciplinary, this doesn't mean that all experts within the team need to have interdisciplinary skill sets, such as: a professional who can program and do design work too. Indeed, team members sometimes do tasks that aren't in their appropriate skill sets. For instance, if the tester is very busy, a developer might jump in and take some of the tester's tasks. This makes it evident that Scrum is all about making the team win! Just think about it in a sport like a soccer/football when Team A is behind 1-0. Team A gets a corner kick, and it is the last minute of the game—then even the goalkeeper will come forward.

The goalkeeper is aware that it isn't the natural position of a goalkeeper, but it is all for the higher purpose of winning.

Furthermore, the Development Team is responsible for ensuring that all work is done. The team should be mature enough to get going after the Product Owner discusses the requirements. Of course, it is sometimes possible that some specialized knowledge isn't present in the team. The team could then search for external help or expertise. These activities remain the responsibility of the team. Don't forget that the team is self-organizing. Another important aspect within Scrum is that we work with fixed teams, which work together sprint after sprint. Therefore, the team grows together, and the team members become more attuned to each other. A good Product Owner needs to know how much time the Product Backlog items—such as requirements—cost. Thus, the Product Owner needs estimates. And who is better to deliver these estimates than the Development Team who have the experience and have to do the work? Do not make the classic mistake of making estimates as a Product Owner or Scrum Master, or leaving it to one of the people in the team. The entire team gives the estimates. I still find it unreal that in some organizations, a manager makes these estimates while she/he has no understanding of the work at all. Fortunately, in Scrum, things are different. By adequately executing Scrum, you gain much more insight into the progress and control of costs. The only thing is that you have to start.

Scrum is a process that is intended for work that has a more complex nature, like product development. These are uncertain ventures, with a clear purpose, but without a clear trajectory to traverse. Due to the dynamism of such projects, planning anything in detail upfront is ridiculous. There are too many variables we can't predict before actually doing the work. This makes in-depth planning unnecessary, to say the least. Many dynamic projects fail, because estimating the tasks is done poorly. Scrum is an empirical process in which we are in a constant "learning mode." Everything we learn is used in the coming sprints and put into practice straight away. Thus, things like designs, programming code, and documentation are always being tailored to the desires and wishes of the customers or end-users.

Once all the work has been planned during the sprint reviews, the Development Team can get started. The team makes the Sprint Backlog by hanging the Product Backlog items and related tasks on the wall or a whiteboard, forming a so-called "Scrum board" by making the sections "To Do," "Doing," or "Done." This gives everyone a good overview of the work in the current sprint, not only for the team but everyone who enters the room and takes a look at the Scrum board. Besides, the progress is tracked by the team in the burndown chart drawn out on another whiteboard or flip chart, and placed next to a note with the team's "definition of done." Thus, the ideal line of progress and how much of the work has been done is indicated, including when we can say the product (increment) is finished.

To further illustrate this, take the following example. Let's say the Development Team works together from Monday through Thursday. Every day starts with a stand-up meeting. This is a quick team consultation first thing when the team members arrive. It is advisable to do it first in the morning, for example, 9 a.m., or 9.30 a.m. if that suits the team better. This is much better than mindlessly checking pending or new emails. Instead, this daily stand-up meeting intends to coordinate work for the coming day. The earlier in the day, the better. It keeps the team members focused so that they can get to work straight away without wasting time. Remember that this is the moment to synchronize and not to report. There are other meetings to report on progress. It is a meeting of the team, just like construction workers on a project have a work meeting in the morning. They would discuss things like, "Oh, this morning, it's likely going to rain, so I'll take care of fixing the roof first, before other tasks." A colleague could say: "Well if you're fixing the roof anyway, I can take care of my task regarding the antenna simultaneously. I'll go with you."

This example falls in line with many of the Scrum values or principles, namely: Dedication, Focus, Openness, Respect, and Courage. These are important to deliver the best possible product; better than the Product Owner might have envisioned. You can always "hack" the system by giving high estimates to things that the Product Owner cannot estimate or verify. How to deal with this will be addressed later, but always realize that you are developing a product for the Product Owner. Try to support it with all your expertise, discipline, focus, creativity, and a sense of responsibility. And if you do not believe in the product, then it is your job to give your say respectfully; be open about it! The Product Owner will only benefit from this in the short—and long—term because this openness or transparency tends to result in better products and more customer satisfaction.

After a week or two, ideally, it's time for a critical moment, namely: the sprint review, wherein the Development Team shows the finished work to the Product Owner and stakeholders. The entire team works toward this moment, to give a demo and receive feedback. The team should always embrace a mindset of growth. When this is the case, the team would rather hear that something is good or not, because this gives them time to readjust and grow in the process to deliver even better products. During the demo, the team can show more than just the screens/webpages or the system. Think about things like test results and documentation. This is useful especially when stakeholders from the management team are present. Instead of a demo, the Development Team can also ask stakeholders to take a seat behind the team's development machines to try the system. Due to this hands-on approach, the stakeholders can get more of a feeling of the system and hopefully give better feedback. Let each team member guide a stakeholder, and make notes of the questions, comments, and especially the feedback. This way, you kill two birds with one stone, by immediately designing a part of the user acceptance test.

After the stakeholders have provided feedback, the Development Team withdraws to reflect on the final period. The sprint retrospective meeting allows the team to take the time to improve. Scrum encourages teams to evaluate the processes and performance of tasks because that is the way to keep growing. It is essential that a team takes the time to review what could be improved and to take a serious look at what is not going well. There are, therefore, no outsiders present at the sprint retrospective. The Scrum Master and Product Owner can join this meeting because both are part of the Scrum team that works toward the common objective.

However, it is vital that all team members feel at ease to share their experiences, missteps, and wins. There are many techniques to craft a retrospective meeting, and this will be addressed later on. For now, you should know that during this meeting, the team members figure out a few points for improvement. Think of things like: "Keeping the stand-up meeting to only fifteen minutes," "Asking an external consultant for advice," or "Making better adjustments to the 'definition of done.'" There is always something that could be better. Especially when the team is new to the world of Scrum, plenty of things go wrong, and that is just logical when starting.

It is the Development Team's responsibility to agree on a couple of concrete actions to improve working with Scrum at the end of the sprint retrospective meeting. Also, committing to take care of these points is necessary—take the following points as food for thought:

- Start the daily stand-up meeting at 9 a.m. and not later.

- Work more in pairs whenever possible.
- Jot down when a team member is absent in a shared file.
- Make more time for the Product Owner to refine the Product Backlog.

In any project, having the right team with the right competencies is essential. But where do we look when we gather our All-Star team? If you are responsible for assembling the Scrum team, you need to know more about how to do this adequately, because the team can make or break any project. So, are you ready to gather your All-Star Scrum team for project success? Let's do this!

Chapter 4: Scrum Teams: Gathering Your All-Star Team

Prioritizing your backlog and working on the items is not a task that is done in solitude. With various stakeholders, the essential things are selected to work on. Before any work takes place, you have to form your "All-Star" Scrum team. The Scrum team is not just a team; it's a multidisciplinary team. Depending on the project, the team should include people with various expertise, such as designers, developers, and business analysts. Every team member is aware of the collaborative nature Scrum is based on. Therefore, no one minds lending a hand and sharing responsibilities for the greater good of achieving the sprint goal. I've learned from experienced Scrum practitioners that they would prefer a "decent" designer, for instance, who fits well in the Scrum team and surrounding culture, than an "outstanding" designer who doesn't. When recruiting someone for the team: make sure they follow the rules; don't mind lending a hand (even if it's for something they are not directly responsible for); and that they fit in the culture.

When you are the one responsible for gathering the team, you should have a few qualities in place. Having these qualities in place is the difference between forming a simply "good" team and a "magnificent" team. These are a few top qualities:

- **Integrity.** Too many people talk about integrity and its importance but don't know what it actually entails. What do we mean when we ensure integrity in a project, or when we work with integrity? The exact definition depends on the context. The context we address here is having integrity in a project environment where Scrum is employed. With this context in mind, the definition mentioned by Cambridge Dictionary fits nicely: "The quality of being honest and having strong moral principles that you refuse to change." Every Scrum Master has to deal with some problems on the team. Having a strong sense of honesty will benefit yourself, but also your teammates. When no one in the Scrum team trusts each other, when people aren't honest, outcomes will be far from desired. The same goes for following the Scrum principles and rules. As a Scrum Master, you should be first and foremost in adhering to these rules and an exemplar in terms of integrity too.

- **Show responsibility.** When you show responsibility as a Scrum

Master, it is easier for the team members to show responsibility for their tasks as well. As a Scrum Master, you have to plan any of the reviews, stand-up meetings, or other ceremonies. If you lack in this area, it will snowball to the Development Team and eventually to the product that the team aims to deliver. I'm a fan of the definition shared by The Nottingham Trent University, which states: "Responsible leadership is about making sustainable business decisions that take into account the interests of all stakeholders, including shareholders, employees, clients, suppliers, the community, the environment, and future generations." I like this particular definition because it makes clear that responsibility is more than self-responsibility. Of course, we need to be responsible for our tasks and finish them with great effort, but there is more to it; responsibility doesn't stop there. By applying Scrum, you will develop a greater sense of responsibility for the product to be delivered, due to the emphasis on collaboration.

- **Be friendly.** I don't know of anyone who would want a manager or Scrum Master who is rude and continuously complains to every person on the team. Instead of being frustrated all the time, be the friendly one. Be the one who brings people up when they are down. Be the one who listens to others and tackles problems as soon as they arise. Be the one who makes the difference between a good and a great team. Behaving positively and being flat-out friendly can change your team's performance for the better, even if the team is filled with beginner Scrum practitioners.

To get more into the process and practicalities: usually, the Scrum team consists of around five-to-nine members. All roles should be represented: Product Owner, Scrum Master, and Development Team. The team should have professionals for all the needed skills to undertake the project. There is not a "one-size-fits-all" way of organizing your team. This depends mainly on the type of project and organization, but there are a few common ways to organize a team.

- **Focus on customers and end-users.** A product being developed during a Scrum project may have a number of variables and numerous types of users. Having an organization based on customers facilitates the development process, because it makes the team focus on what the customer actually wants and not what the team *thinks* the customer wants.

- **Focus on products.** These types of teams are usually seen at start-ups because they don't have the amount of complexity larger companies have. Thus, products can be developed with less-complex features at a faster pace.
- **Focus on features.** When working on a project, the team wants to focus on adding, removing, or innovating features. This is especially useful if the product to be developed is too significant for a single team to move forward.
- **Focus on a combination of factors.** Though there are various ways to organize your team, it may well be advisable for your team to take the best of these mentioned **ways** of organizing. Thus, a combination of focusing on the customer, products, and features, are most useful in practice. Otherwise, the team might focus too much on a particular subset, while the project is interconnected between multiple subgroups, such as the different customers, products or types of products they desire, and specific features for each type.

Besides the sprint planning, we should take a look at how the Development Team members work together. Within Scrum, we work with multidisciplinary teams with various expertise. Let's say we have two user experience (UX) designers, three software developers, and one tester. The UX designers focus on the look of the product increment; they make various mockups, wireframes, and more. The software developers write the code for the product increment, and the testers make sure test cases are taken care of. As you may consider, this can still create various boundaries, as we've seen in traditional methodologies. These boundaries may be physical but also organizational. What if the designers want to keep on designing until everything is "perfect?" And what if the developers want to write all code "perfectly" before they discuss it with the tester? Because they're not finished with the tasks, they sit in separate rooms. This is far from ideal. Therefore, make sure you take care of the following to avoid the pitfalls of traditional methodologies:

- **Make sure the professionals sit together.** You all are a team for a reason. A team is meant to be close to each other and not in separate rooms. Let the team members sit with the professionals they work with in the sprint. Place their desks next to each other and remove the physical boundaries as far as you are capable.

○ **Don't wait.** Please, make sure not a single team member is waiting at all. Instead, *embrace the iterative-way of working* proclaimed in agile methodologies. For instance, for the designers, show early sketches to the developers. And developers, when you are done with the first half page of the product increment, show it to the tester and explain the details. Afterward, the testers can figure out a proper way of testing the product increment, interacting with the developers for clarity. For example, testers could think about: "What happens if the customer adds an 'and' in the search function?" More of this will be discussed later, because these types of waste are crucial in Scrum projects.

○ **Limit "work in progress."** Like the previous point, this isn't necessarily something within Scrum but more from Lean and Kanban. Despite this, it can be a factor that changes the overall outcomes of your Scrum project. Thus, make sure you limit "work in progress" concerning the various team members. For instance, if the software developers pound through and finish three pages, but the tester can only test one page, this creates a bottleneck in the process and leaves the tester overwhelmed. With a team that embraces iterative and collaborative working, this is easier than you might think. Work on one thing, e.g., one page, finish it, and go to the next thing.

When the Scrum team works, it is essential to have a continuous flow day in, day out. UX designers and software developers are designing and developing the software. The developers are writing a unit test and testing what they are doing, to a certain degree. Then, this work is checked into a source control that is hooked up to continuous integration. Continuous integration helps the team to design quality software faster. Also, it helps deliver new functionality to the customers or stakeholders more quickly while the developers become more productive and improve the quality of the software. With the continuous integration tools such as the ones created by Google Cloud, you can create automated builds, perform tests, deliver environments, and scan artifacts for security vulnerabilities, all in a matter of minutes. Continuous integration is essential for making your team work like "one body." In short, here are a few benefits:

- **More efficient development and improved productivity.** Accelerate developer feedback by running builds and tests on machines that are linked via high-performance networks. Perform builds in parallel on multiple computers for rapid feedback. This results in spending less time detecting errors.

- **Scale to the moon without worrying about maintenance.** Are you worried about the lengthy design and test times your team faces when it grows? There are various tools for continuous integration that scale automatically. This allows the team to make a hundred or maybe even thousands of builds when the amount of team members or projects grows.

- **Make secure incremental products part of your team's efforts.** Don't figure out security at the end of a sprint. Make sure security is checked continuously. If the team members don't want to take care of this on their own, make sure you get some tools in place to do that. Many tools can scan for security issues as soon as new artifacts are introduced. They even give the option to export detailed reports on the impact of these security issues and possible solutions. Besides, it's possible to set policies for different working environments. Thus, only verified artifacts would have a place in the end-product.

- **Grant your team more flexibility.** With continuous integration software, you can pack your source code in Docker containers, for instance, or non-container artifacts. This makes it possible to build tools that we see in many organizations, be they small or large, such as Maven or Go.

When a new code is checked into this system of continuous integration, it immediately picks it up and builds it with everything else. This should be understood in a team that frequently works in the same code base, such as Go. When this is the case, the second a developer checks something into the source control; she/he knows if it makes anything break that someone else developed. Afterward, you can focus on running feature testing, such as the search feature. Usually, in preliminary stages, the testing will be manually done by team members.

For customer satisfaction, testing is imperative. It is crucial for our customers that the product the team develops works well. That means that there are no errors, bugs, and that the product does what it should do based on the set requirements. In Scrum, the developers test whether what they have developed works, but for large projects, this is a lot more complex and time-consuming. A simple search feature form will, in some instances, quickly have dozens of different outcomes, or scenarios, based on the form. All these scenarios must be checked. Fortunately, we can automatically test these types of scenarios using various tools or software. Although automating testing is great, it is not always possible. A disadvantage of automating this process is that setting up automatic tests takes time and money. Clear agreements with the team must be made about which scenarios are and are not being tested, how often they are tested, and what a test looks like. Tests must then be written and looked at by various experts working on the product. Many types of tests can be automated, but with a user test, this is a bit more complicated. Manual testing by the target group is usually essential here.

On the other hand, there is tooling to measure user experience automatically. Automatic testing can provide a sense of false safety. Nevertheless, it is essential to monitor the tests and the product increment the team is working on. It could be a feature of a website, application, or webshop. Thus, automating tests doesn't mean that no team member has to do anything to get this up and running. However, when everything is up and running, it will have a tremendous effect on the team's productivity in the long term.

What you have to remember is that testing is not something done at the end of the sprint! Testing should be done (nearly) every single day the team gets together. Continuously testing makes possible bottlenecks visible and allows the team to adjust rapidly, whenever necessary. This is the concept of inspecting and adapting at a micro level, namely within a sprint.

At first, if the team members are new to Scrum, they might see working in these small increments as less efficient. Such as when a developer says, "I can develop more pages, why do I have to wait?" Well, the developer may be able to develop ten things, but we can test only three. Developing more is not useful, because we cannot release untested code. Later down the line, this approach of sometimes doing less will be worthwhile, because nothing—or very little—has to change to make things work at the end of the sprint. Letting every professional take care of her/his work in its entirety before discussing was a critical element in how projects were tackled in the past. However, this resulted in many frustrations, because a lot had to change further down the line. With Scrum, these changes can be taken care of immediately after the professionals have discussed their progress. Therefore, in a later stage, there's not much to "clean up" or repair, which saves precious time and money.

Chapter 5: Scrum Artifacts

Besides particular roles, Scrum has multiple artifacts. In practice, we see that Scrum teams start with a so-called "product vision." Even though this is not necessarily part of the Scrum Guide, I've noticed that projects struggle without one. Without this artifact, it is difficult—maybe even impossible—to move forward in a project and garner excellent results. The product vision helps us get on course concerning dealing with the project. It makes clear who our target market is, who are the people who will need what we produce during the project, and what their desires are. What challenges are they facing? Also, the product vision should be clear about the specific business need or opportunity the team is going after.

Furthermore, it describes the key elements that are necessary for the product that will be developed during the project. Lastly, there should be a thorough understanding of what value the project will deliver to the organization. Usually, the "why" is vague for people working on the project, making it difficult to work hard toward the goal. The "why" could be anything from the amount of money made for the company to creating more impact for your specified market.

Another critical artifact that may not be part of the Scrum Guide is the release plan. Although it is not necessarily part of the Scrum Guide itself, in any project we have to know the game plan, and the release plan eases this process. Answers to questions like: "How are we going to tackle challenges and overcome unknown obstacles?" and "When will things be delivered?" are forecast entirely based on empirical data. Empirical data is data on how the team has performed in the past. So, it's not about how we think we are going to do it, but what the team members have proven that they *can* do. You should know that the release plan has an overlay on top of the Product Backlog. It tells us how many things of the Product Backlog we can get done in every short feedback loop. Also, the release plan is updated every sprint because while we work, we gather more and more empirical data on how the work is going, and if the team is on the right track. Below you'll read more about the most crucial Scrum artifacts found in the Scrum Guide.

Product Backlog

Probably the most crucial Scrum artifact is the Product Backlog. This is one of the two primary lists used in Scrum. This is the artifact that is managed by the Product Owner, as explained earlier. Why is this artifact so key? It is vital because it's the source of all the things that are required in the incremental product. It is a well-ordered and well-prioritized list of all functions, requirements, fixes, and enhancements for the incremental product. There is no other artifact or document where requirements are listed. All team members refer to this artifact, and this one only; there shouldn't be multiple versions going around.

The Product Backlog is the list that contains everything that needs to be done to create the product. It is that simple, and things shouldn't be complicated for no reason. The idea is that you know what to do, and that everyone knows what has been agreed upon, but that the work still has to be done. Scrum is not just starting without a goal or direction; as the product vision shows. But, in Scrum, we recognize the fact that things change during the work process and we constantly evolve as a team. That is why the Product Backlog is never a static list of items. New insights mean that new things will be added to the Product Backlog; earlier issues will be removed if they have been tackled; and the order of the items can change around. The development of new products is too complicated to realize with a preconceived plan. The Product Backlog is far from being a preconceived plan, it makes room for the innovative and dynamic projects we face nowadays. The critical point is that the Product Backlog only answers the "what," i.e., the properties of the product in a functional sense. For example, the function "Add to cart." The goal of the Product Backlog is not to explain the "how," for instance, how the Development Team should go about figuring out the "Add to cart" backlog item. That is the team's job to figure out.

By limiting yourself to answering the "what," everyone involved stays on board. This backlog is made for everyone: the team, but also internal and external stakeholders. Thus, it is imperative to make it understandable for every party involved, be they technical or non-technical. Although the "how" is not answered in the Product Backlog, it is good to have an explanation of "why" a particular item is listed on it. When the backlog is clear and free of jargon, it makes for a very efficient communication tool and entices stakeholders to engage in meaningful discussions.

The Product Backlog is managed by the Product Owner. The Product Owner is the boss of the Product Backlog. Only the Product Owner can determine whether something takes place on the backlog and with what priority. As a team member, never use your possible technological know-how, or whatever know-how you may have, to get things on the Product Backlog. In other words: as a Product Owner, don't include items on the backlog that come from the team that you don't understand. If they explain it and it becomes clear, you can always add it to the backlog. However, if you are in doubt, do not add it to the backlog. Usually, you can identify these questionable items, when a team member uses words such as "generic," "regret," and "later," in her/his explanation.

I hope you've figured out by now that the items on the Product Backlog are called Product Backlog items. Yup, not very innovative, but it does the job. What we usually see in real Scrum projects is that these Product Backlog items are formulated in user stories. User stories are a particular way of describing functional requirements on the Product Backlog. These stories aim to specifically express the functional wishes of users. This will be explained later on. However, not everything on the backlog is a user story or requirement. Various types of items can be part of the Product Backlog, such as:

- **Problems**. During the project, multiple bugs or issues can appear that need to be fixed to move forward. So, don't create any other spreadsheet or document with bugs or the like. These should have a place on the backlog.

- **Requirements**. Of course, requirements have a place on the backlog too. These can be of numerous kinds, such as functional requirements, non-functional requirements, system requirements, et cetera.

- **Desires and wishes.** When working on the product, stakeholders can deliver feedback and share their insights. The same is true for customers, and they can desire certain features. If the team sees the value in these matters, they can be added to the Product Backlog.

When these types of items have a place on the backlog, this creates more transparency. After the backlog items are defined, it is time to prioritize them. The essential backlog items need to have a place at the top, and they need to be in order. To facilitate this process, you can make use of the MoSCoW method, which is a method to prioritize tasks and come to a common understanding with the team members, (which job should be dealt with first, which one after that, et cetera.) MoSCoW is an acronym for the following:

- **Must-haves**. These are the requirements that have to be in the end-product. Without these, the product is useless.

- **Should have.** When the "must-haves" are taken care of, these should be tackled next. These are the matters that are very desirable, but the product could survive without them.
- **Could have.** These are requirements the team should only embark on when there is time remaining in a sprint.
- **Won't have.** These are the matters that won't be taken on in the current sprint(s) but may be useful in future projects.

Using this technique is useful, but make sure the team doesn't fall into some common pitfalls. What can happen is that team members will put all—or too many—items in the "Must have" category. Always double or triple check if these things are vital to bring the product to life. Furthermore, team members can be biased when the items are categorized together. Thus, you could let everyone categorize the items themselves first and then discuss each team member's categorization. This leaves room for discussion and brings about valuable insights. As a team, we should dare to come up with simple solutions that leave space for adding complexity in later stages of production. Figuring out simple solutions can help to deal with tackling the backlog items, because larger projects can have around 55-65 items. In practice, you could even come across an organization with hundreds of backlog items. In short, this is far from ideal, and more than likely the Product Backlog is far from optimized. Always reassess if the backlog can be improved upon and readjusted. If the backlog seems to contain too many tasks, see if it is possible to combine similar jobs or reduce the number of tasks if they aren't necessary at the moment. Doing so will keep the backlog clean and clear. Even this type of clustering can be done in silence by the team members to avoid possible biases.

	A	B	C	D
1	**Priority**	**Estimate**	**Description**	**Remark**
2	200	8	As a vacation shopper I want to compare different types of transport so that I can	
3	400	2	As a vacation shopper I want to receive a summary of my booking in my e-mail.	Make sure a summary appears on the webpage after booking and an email gets sent instantly afterwards.
4	1200	4	As an administrator I want to generate and track affiliate links	

Example initial Product Backlog

Furthermore, you should note that more than 85 percent are functional requirements. If you spot a lot of technical requirements or bugs on the backlog, this should alarm you that there are some serious issues with the quality of the product. If you're not sure if a backlog item should be removed, a cool technique to help you get more clarity is called "five times why." This technique enables you to figure out the root cause or underlying problem of the backlog item by asking "why" five times. Thus, you can evaluate if it makes sense to leave it on the backlog or not. Below is an example of the method.

Let's say there is a backlog item concerning a customer who is receiving orders late three times in a row.

- **Why Number 1:** Why did the customer receive these orders too late?

- o Answer: The transport company, responsible for the delivery, did not have the correct address details of the customer.
- **Why Number 2:** Why does the transport company not have the correct customer address details?
  - o Answer: The address on the shipment does not match the address of the customer.
- **Why Number 3:** Why does the shipping address not match the customer's address?
  - o Answer: The customer moved to a new location three months ago, and this new address is not yet included in the supplier's customer database.
- **Why Number 4:** Why is the customer's new address not yet included in the supplier's customer database?
  - o Answer: The administrator forgot about it and has been ill now for a couple of weeks.
- **Why Number 5:** Why didn't anyone think about changing the address?

  - o Answer: No one besides the administrator has the right to make a change to the customer database.

Furthermore, there is a technique for sorting the Product Backlog items by having stakeholders vote on the items regarding their importance. For instance, this can be done by giving everyone two to three votes and having them distribute these over the clustered backlog items by placing a dot next to them. If the votes are divided and the order becomes apparent, as Product Owner try and get rid of as many items as possible with the lowest or no votes. Just think about it: if there are forty votes to be distributed among twenty clusters of backlog items, then you can safely throw away every item with two or fewer votes. You can use the same technique with stakeholders to elicit more responses.

Seriously, the power of just eliminating stuff is far underestimated. Nothing is easier than not implementing items: it takes no time, has no bugs, no maintenance, and no documentation! Pay attention to the stakeholders' emotions. Some items with few votes can be essential for a stakeholder, so if you're thinking of throwing an item away, before you do have a good discussion. It is strange that someone finds an item very important, but that nobody else seems to care. Perhaps there is no support, and the stakeholder must accept that. Probably there is no understanding, and the stakeholder must create support. As a Product Owner, you can encourage your stakeholders to give their say about the decision they made. If this brings up issues you can't solve, then you know some issues need further investigation. This makes you aware of probable obstacles right off the bat, which is much better than learning about these problems when it's too late in the process. To illustrate this further, it may be that a particular stakeholder is much more important than the average stakeholder. Give them more opportunities to have a say so that it is clear to other stakeholders their opinion has a significant impact. In the coming chapters, you'll learn more about the practicalities of the Product Backlog, such as making proper estimates and crafting user stories. However, for this chapter, we'll take a look at a few more critical artifacts.

Sprint Backlog

The second crucial Scrum artifact is the so-called Sprint Backlog. This artifact is derived from the Product Backlog and can be seen as a plan for delivering the backlog items for each short feedback loop. The Sprint Backlog contains all the items for the sprint the team is currently working on. In the Sprint Backlog, we find all the tasks to fulfill a backlog item. For example, take the backlog item: "Banner Area" for a web design agency building an e-commerce website. The user story for this item is: "As a marketing professional, I want to be able to make an advertisement so that I can get customers for our products." Corresponding tasks that will appear on the Sprint Backlog are: "Make a banner area on the website," "Give the marketing professional the right to place a banner on the website," "Test if the banner is available for customers."

Furthermore, the Sprint Backlog is an artifact owned by the Development Team and not by you as a Scrum Master or Product Owner. The team comes up with it, and they manage it and keep it up to date. Of course, the Sprint Backlog is dynamic and should be made readily available and visible.

At the start of the sprint, the Sprint Backlog is created during sprint planning. The most logical thing is to take the top items on the Product Backlog and use these as the basis for the Sprint Backlog. But it may be useful to choose a slightly different composition depending on the goal you want to achieve during the sprint. This is determined in the first part of the sprint planning meeting. It may be that the Product Owner links a number of items together that form a theme, so that the Development Team can take the product into production at the end of the sprint. It may also be that based on the latest sprint review, other choices are made, rather than simply picking up the items at the top of the Product Backlog.

The velocity of the team determines the number of items that are picked up; that is the number of points (story points) that a team can tackle in a sprint, and this is typically based on the results achieved in the last sprint (more on this later). It may, of course, be that a team member is absent or that someone is on training. Then it is good to adjust the velocity somewhat.

When filling the Sprint Backlog, be realistic and do not take on more tasks than you think you can handle. Especially after a disappointing sprint, inexperienced teams sometimes want to gain some confidence by taking up fewer items. Choose as much work as you can deliver based on recent results. In the second part of the sprint planning, the team will go through the items in more detail and break them up into more technical parts, so that it becomes clearer what exactly needs to be done. This is the time when the items on the Product Backlog, which mainly describes the "what" and "why", are translated into tasks that describe the "how." These more detailed tasks, together with the original Product Backlog items, make the Sprint Backlog.

The Sprint Backlog is created by—and for—the Development Team, as a tool to make the division of work clear and to monitor progress during the sprint. To be more transparent, the team generally shares the Sprint Backlog with everyone. The best way to do this is to hang the Sprint Backlog on the wall of the team room or to write it on a whiteboard to create a "Scrum board" (this is a board with the Product Backlog items you take on in a specific sprint and the tasks extracted from each item). Use markers, paper, and Post-it Notes to create a Sprint Backlog: it's that easy. You can use colors for various types of items, for example: green for Product Backlog items, red for bugs, et cetera. You can also give tasks-in-progress a label with the name of the people who are working on them. Make the board your own, but don't forget to add the fundamental Scrum elements first.

It is up to the team to determine the level of detail that it needs to complete the tasks. When the team has discussed the Product Backlog items and the

Sprint Backlog is ready, the work can begin. However, always make sure the Development Team has made a substantial commitment to the Product Owner.

Burndown Chart, Impediment List, and the Definition of Done

Two other important Scrum artifacts are the burndown chart and impediment list, which track the work that remains each day. The burndown chart provides a visual of how many hours are remaining to deliver the incremental product of the sprint. If you are updating the Sprint Backlog, there isn't much difficulty in updating the burndown chart for the Development Team as well. The burndown chart is purely to get a quick view of how the team is performing and to see if it is on track timewise. The burndown chart can be displayed on a whiteboard or digitally with a flat-screen. The essence is that, like the Sprint Backlog, it is readily available and visible. No team member should be able to overlook it.

The next Scrum artifact is the necessary to be visible impediment list. For this list, we jot down anything and everything that could block or affect the route and slow down achieving the objective. The Scrum team should update this. If you are the Scrum Master monitoring the team's progress, this artifact is crucial. You could add to it, but the Development Team should do this first and foremost. If something can't be fixed at the moment, this can be escalated to the Scrum Master, who will find other professionals to take a look. However, this doesn't mean the Scrum Master owns it.

In every sprint, it is essential to have a "definition of done" for the team. In previous chapters, we've spoken about the definition of done. It is not exactly a list, but it is related to the lists. It answers the question: What do we mean when we consider that the work is "finished?" Agreements have to be made beforehand to decide what finished should be. This can vary wildly from project to project, but the standard is that "ready" in Scrum means that you can take it to production. The definition of done is crafted by the Product Owner and Development Team. For instance, let's say a fictional software company called xSoft Solutions wants to make a "definition of done." The Scrum team should consider these components:

- All the code is written and meets the agreed standards.
- The work has been functionally tested. Does everything work as it should for end-users?
- There is documentation available when needed to explain certain

parts or features.

- Et cetera.

The definition of done can be jotted down on a piece of paper and placed next to the Sprint Backlog or Scrum board.

Chapter 6: Scrum Ceremonies, Meetings and Agendas

Within Scrum, there are various events or ceremonies. Most people don't find difficulties in attaining knowledge about Scrum, but do so when implementing the knowledge they garnered. This chapter demystifies the execution process of Scrum, by looking at the sprint planning meeting, working as a team, holding the stand-up meeting, quality control, and how to groom your Product Backlog.

First, as previously discussed, we start things off with the sprint planning. Selecting the items to work on should be done with the knowledge of past performance and capacity. If team A, for example, has been consistent in delivering four backlog items each sprint, this can be taken into account. The same goes for capacity. If, for example, a team member becomes sick, this void should be taken into account. Make clear *how* the items will be delivered in the Sprint Backlog: all the tasks you need to do, such as documentation, testing, design—whatever it is to finish the product increment.

The Sprint Planning Meeting: Execution

The complete Scrum team will be present for the sprint planning meeting: The Product Owner, Development Team, and Scrum Master. Besides factoring in the team's speed when executing backlog items, team capacity should not be overlooked. Will all team members be present during the sprint? Is a team member going on vacation? Will the whole team visit a seminar? Et cetera. A backlog item at the top may be switched with another less intensive backlog item that fits well with the available capacity. No matter what the capacity, having a reasonable pace to tackle the backlog items is vital. The team should only bring in the items they feel they are most likely to complete entirely.

Furthermore, you should determine your sprint length. The length varies on the project and availability of team members. You shouldn't go shorter than one week and not more than four weeks. Once you have determined the length, stick to it. Bring more items in if the sprint was finished earlier. With the length in place, identify and jot down a sprint goal. This goal has the character of a mini product vision. It will give a general idea of what product increment the Development Team is working on.

To further illustrate this, take the fictional e-commerce company: Pineapples Inc., which is working on a new website. When the Pineapples Inc. team held the sprint planning meeting, they decided to take on the following tasks from the Product Backlog:

- Add "search the catalog" for customers to find products to add to their online shopping carts.
- Provide valuable product suggestions to repeat customers.
- Update and expand the payment platform.

If the team finds that the velocity and capacity doesn't fit the selected items, an item could be switched to match these. Fortunately, in our example, the Pineapples Inc. team has done a great job picking the tasks so they don't need to be switched out. Thus, based on these tasks, the sprint goal should be formulated. Now, this may sound like a chore, since the items mentioned above seem diverse. However, the Product Owner didn't prioritize these for no reason. She/he most likely values the aggregate contribution of these three items and has an image of the product in mind when receiving an increment combining these backlog items. The sprint goal could be: "Develop a self-service, transactional solution that allows Pineapples Inc. shoppers to buy goods and receive product suggestions."

When you create the Sprint Backlog, the Product Owner and Development Team will gather in the same room. The Product Owner will then present the backlog items that they will take on, answer any issues, and then the team will discuss the design. For example, let's take the first Product Backlog item mentioned above. This is all about "searching the catalog." This backlog item can be divided into smaller tasks, such as creating the actual page where customers can search, writing the code and queries, and testing the search function. After the tasks are defined, place the expected amount of time each task will take. For example, creating the page will take seven hours, writing the code and queries will take another ten hours, and testing everything costs eleven hours. All elements needed to complete the backlog item should be reduced to tasks like these that can be taken on day in, day out. Of course, the hours spent can vary and can change as time progresses. But don't worry. Eventually, the team gets smarter and more accurate at assigning times to specific tasks.

Successful Scrum teams tend to use a primary method when creating these smaller tasks, or Sprint Backlog items. With ample focus, they strive toward tackling these items even if they seem challenging. Everything the Scrum team does—every decision they make and action they take—is done with an acronym in mind, namely: SMART. Intense concentration on fulfilling the elements of this acronym is what enables excellent Scrum teams to accomplish what other teams would only see in their dreams. According to a study from Willis Towers Watson, half of all managers do not set practical employee goals, let alone goals for Sprint Backlog items. So, what does SMART even mean, and how can it help the Scrum team?

SMART is a way to check whether objectives are **S**pecific, **M**easurable, **A**cceptable, **R**ealistic, and **T**ime-bound. Goals are often formulated vaguely. They seem more like wishes than concrete goals. To finish a sprint in the stipulated time, I highly advise you use SMART.

Specific. Being specific means that it is clear to everyone what the objective is and what result you want to achieve. To make a goal specific, ask these questions:

- What do we want to achieve?
- Why do we want to achieve it?
- When does it happen?
- Who is involved?
- Where will it happen?

In brief, describe the goal clearly and concretely, with a perceptible action, behavior, or result to which a number, amount, percentage, or other quantitative data is attached.

Measurable. This relates to the quality of the efforts to be made. How much are we going to do? How can we measure that? What is the result when we are finished? You must be able to see, hear, taste, smell, or feel a SMART goal. There must be a system, method, and procedure to determine the extent to which the goal has been achieved at a given moment. If possible, conduct a baseline assessment to determine the starting point.

Acceptable. If you set a SMART goal for yourself, it is enough that you accept it yourself. However, when you set a goal for a group of people, there must be support and the team members must agree with it. Otherwise, they will not take the necessary responsibility to achieve the goal. When individual objectives and organizational objectives are not aligned, the goal will not be delivered, or the change will not last. There are various methods to make sure goals are well-aligned. The primary way is to make sure there is engagement between you and your team members. You have to actively involve your team members in choosing and formulating the objectives. Every team member must have an opportunity to give her/his word.

Furthermore, some experts tend to explain "A" as: "Action-oriented" or "Achievable." These terms indicate necessary elements for a successful goal, namely, they show us that a goal needs to provoke action and be something the team can achieve. Keep in mind that a goal formulated the SMART-way prescribes a particular result, not an effort.

Realistic. Is the goal achievable? Is there a feasible plan requiring acceptable effort? Can the parties involved influence the requested results? Do they have sufficient know-how, capacity, resources, and capabilities? It is essential to take a closer look at these aspects because an unattainable goal does not motivate people. Sometimes the "R" in SMART is also explained as "Relevant." A feasible and meaningful goal is motivating. A realistic objective takes practice into account. There is no organization where people work on one goal for one hundred percent of the time. There are always other activities, unexpected events, and distractions.

An objective can also be unrealistic if it is imposed on the organization at a level that is too low. For instance, the goal: "Increase profits by nine percent in one year," is not a good target for the marketing department, because profit is an integral result of the entire company. A goal that is too easy to achieve is not exciting either, because it does not challenge the Scrum team. It is best to set goals that are just above the level of the team so that they keep improving. If people feel that they have to go the extra mile to hit a goal, they'll feel much better when they achieve it. This fosters energy for further goals in different sprints.

Time-bound. In general, SMART is used for more short-term goals. Thus, it's perfect for making more sense of the Sprint Backlog items, because these are carried out daily. Take into account that it is essential to know that a SMART objective has a definite start and end date. The following questions can help you further:

- When will we be ready?
- When do we start the activities?
- When has the goal been achieved?

To further illustrate a well-defined and not-so-well-defined SMART-goal, check this out:

> o Well-defined: "At the end of sprint one, on Monday the second of December, the team wants a finished search page with features X, Y, and Z, where customers can search the catalog to find products to add to the online shopping cart."
>
> o Not-so-well-defined: "As a team, we want a nice search page for customers." This goal is not specific enough, not measurable, not acceptable, not realistic (because you don't know what to do), and not time-bound.

When the goals are set, make sure everyone commits verbally to the goals or Sprint Backlog items.

We've previously mentioned sprints, but let's zoom in to know what the Sprint event is all about. In most cases, the sprint is time-boxed to a couple of weeks, anywhere from two-to-four weeks and nothing more. Don't waver from this unless there is a great reason to do so.

As you know, setting goals is crucial for any success. The same goes for a successful sprint. Therefore, make sure every sprint you run has an understandably stated sprint goal, visible for anyone on the Development Team and any stakeholder. After each sprint, you should have a releasable increment of the product. For example, a login screen and the ability for users to log in to the platform of an e-learning company. Though the login screen may be done, this doesn't mean you deliver this incremental product to the market. But it should have the potential to go to production with as little effort as possible. Finally, the scope is set by the Scrum Team and not the sales or marketing department.

Stand-up Meeting: Execution

Another event is the Daily Scrum or stand-up meeting. This event is time-boxed at not more than fifteen minutes. With a Scrum team of around five-to-nine members and a concise Sprint Backlog, this is long enough. Beware of exceeding this stipulated time. The stand-up meeting is not the time for intricate details in terms of design or development. You can have other meetings to do that. Specifically, with this event, you get to know what work every team member did yesterday, what they will do today, and if they have encountered any obstacles along the way. During the stand-up meeting, each member of the Scrum team answers the following questions:

- What have I achieved since the last stand-up meeting?
- What am I going to achieve today?
- Do I expect obstacles, and can the team help me in some way?

Some people may "answer" the questions without answering them, so it is important to be clear and give context. In my experience, I have heard various professionals answer these questions, but many other team members were still oblivious to what they were doing.

There are many ways to conduct stand-up meetings; some may be more constructive than others. Holding the daily scrum or stand-up meeting is something you do every working day. The stand-up meeting is primarily for the Development Team. The Scrum Master should be there as much as she/he can to make sure everything is going well. If there are other stakeholders available, they can come, but they should keep any questions until after the meeting. Make sure you make a place for the stand-up meeting; keep the same time and place for this meeting.

To make things clearer, fill your physical Scrum board or digital Scrum board. This Scrum board gives the team a visual representation of the work. The team members will still answer the questions, but it now becomes more visible. Thus, even if you, as a team member, might not be working on a task, you can see the tasks that aren't done yet and if anything is blocking a task from being performed.

An example of a Scrum board is the following: a white board divided into five sections with Post-it Notes attached to each section. The first section is called "PB Items." this is all about the Product Backlog items you choose in this specific sprint. Again, this can be anything from user stories, all the way to bugs. Usually, user stories or bugs are grouped to take care of them in one sprint. However, sometimes, this is not possible due to a pressing bug that must be fixed as soon as possible. Thus, you'll end up with a few user stories and a bug to fix in the first section of "PB Items."

Afterward, jot down each task on a Post-it Note and place it on the Scrum board. The second section, "Not Done" indicates the tasks that are not yet completed. This is where all the tasks start. When someone from the Development Team takes on a task she/he moves the Post-it to the third section, "In Progress." This section contains all Post-it Notes for tasks that people are still working on. Afterward, in the fourth section, "Done," we place all the tasks that are finished. If all tasks relating to a specific user story have been finished, for example, we then put that user story in this section as well. During the sprint, some impediments may occur. Therefore, the fifth section is designated to deal with obstacles the team may encounter. This way, the Scrum Master can quickly see the barriers blocking the team from moving forward, and figure out a solution, so that the team can continue with the tasks for that particular backlog item.

It is too bad that in practice, we see that a lot of impediments are not tackled. The Scrum Master should bring these forth and get rid of them. This is especially the case if the same obstacles pop up for several days. Even if it is the Scrum Master that solves the obstacles, the team should make them clear, because they are working on the tasks where barriers occur. To give the team a better perspective on the tasks and time they take, we make a burndown chart. The burndown chart provides the visual information needed to manage your project or Scrum sprint daily. The graph shows the remaining amount of work left for the total project or the current sprint. This progress is made clear with the help of two lines:

- Line 1: Remaining work.
- Line 2: Ideal situation.

The sum of all the points on the Product Backlog is the total amount of "work." In other words: the whole estimated "size" of your project. You make the progress of the project evident by continually comparing the number of delivered story points to the elapsed time, marking this on the burndown chart. Most of the time, this is done to cover the number of weeks a sprint has.

Sprint Review and Sprint Retrospective

Finally, we'll take a look at the sprint review and sprint retrospective. The sprint review is when the Scrum team shows the work done in the sprint to the stakeholders, to collect feedback. This meeting is also referred to as "the demo," but the sprint review isn't just about that. Primarily, it is about the feedback and what you do with it. The sprint review takes a maximum of two hours for sprints of two weeks. The sprint review is a meeting where mutual understanding is cultivated, because it is never the case that you can describe exactly what you need in a product well in advance. Besides, most desires and wishes change during product development, and traditionally more features are required than what was initially thought. Therefore, it is important to get as much feedback as possible for every sprint.

It is costly to make something and only hear about the changes needed very late in the process. In Scrum, we do things differently, because we know that the customer changes her/his mind constantly. Scrum gives a lot of freedom, but this only works with a lot of discipline, hard work, and communication.

In this meeting, the team gives a demo of the completed work. That is often a demo of the application, but that can also be other things. If the product increment isn't done, doing a demo is not a very smart thing to do, but it is important to mention why things aren't completed. For instance, if the application does not have a user interface yet, the team can also show test results and documentation. An All-Star Scrum Team can do a demo at any time in the sprint. Always try to get under the skin of the stakeholder and try to make the demo as interactive as possible. Do not show results that do not meet the team's definition of done. You want to entice the stakeholders to push the Product Owner for a release at the end of the sprint. If things are correct, you could receive questions like: "Why don't we put this into production?" It would be painful to admit that the product is not ready because the definition of done contains an error, like something that can't be accomplished in this given sprint.

The sprint retrospective is all about the Scrum team inspecting, adapting, and looking back at the sprint. Here we talk about the positives and negatives. What elements or tasks went well? What items or tasks didn't go well? What can we do better in the next sprints? This is the event where the Scrum team should learn multiple lessons. These can be jotted down in a plan to improve after each sprint. This continuous improvement is what sets Scrum apart from other methodologies. A good sprint retrospective stands or falls with its atmosphere. Ideally, it should have an atmosphere in which team members feel at ease and that they are safe to give and receive constructive criticism and discuss errors. If the team can't do that, it becomes challenging to find improvements, and the observations remain superficial. That is why it is especially visible in a sprint retrospective if a Scrum Master is also the manager—or behaves accordingly. A team that feels assessed by the Scrum Master, whether that is literal or not, is less likely to speak up.

Many elements can be named for creating a winning environment, such as making evident that the meeting is for the improvement of the entire Development Team. It is an excellent opportunity to stand still and look back at the sprint, to come up with some worthwhile improvements. Usually, the Scrum Master facilitates this meeting. It is precisely the way of Scrum to challenge yourself to become better continually, and that is why this is often one of the most important meetings for the Scrum Master. Not only to facilitate it, but also to challenge the team to always keep improving themselves not only as professionals, but also as people. Furthermore, the Product Owner could join this meeting as well. If the Development Team doesn't feel this is the right thing, the retrospective can be divided into two parts; one with the Product Owner and one without.

After the meeting, the release plan can be reviewed and updated based on the empirical data, such as elicited feedback from stakeholders. When the sprint retrospective is finished, the entire process is repeated by starting from the beginning, i.e., with the Product Backlog and the sprint planning sessions. In the coming chapter, we'll delve deeper into how all these things tie together!

Chapter 7: Breaking Down a Scrum Project

In the previous chapters, we learned more about Scrum and its roles, artifacts, ceremonies, and more critical concepts. But how are all these matters implemented in your organization? What does Scrum look like in practice? These and many more questions will be addressed in this chapter. I'll give you insights into how you can assemble the Scrum team, create your Product Backlog, user stories, prioritizing items, estimation, the release plan, and how all these are supported by "Sprint Zero" and a product vision.

Sprint Zero and Product Vision

When the team wants to embark on their project journey, usually, there is a lot to "fix" before the team is ready to go. Just think about crafting your product vision, setting up the initial Product Backlog, and filling and prioritizing the backlog with tasks for at least two sprints. We take two sprints because this keeps the Product Backlog concise enough, but also leaves room to take on additional tasks when there is room for it. Although you might wish to start straight away, you should take care of these things first. A way to "begin" the project straight away and still prepare what is needed, we introduce Sprint Zero. This is the sprint where we set everything up, tidy up things, and make the team ready to fly. Thus, sprint zero is also great for crafting your initial release plan. And don't worry about not having sufficient empirical evidence to start crafting this plan or the backlog. Just make sure something is in place and understand that all artifacts are dynamic. Furthermore, this is the sprint to create an environment for success. Besides getting your team members focused on the sprint goal, having an environment that favors continuous integration will come in handy.

For instance, if a software company were to start sprint zero, they would create or set up a place where the coding rules are clarified; where the programmers have to write the code; where the code is tested; and where it can be deployed. Having such an environment in place allows for seamless and quick processes during the entire sprint. With each sprint, we aim to deploy something useful as soon as possible. Don't get me wrong; this "something" doesn't have to be unusual or perfect, ready for a customer to delve into. Nope, not at all. But it should be something you can move forward with and which complements other product increments later down the road. Undertaking a "Sprint Zero" is not imperative for your Scrum project, but it can help you set things up before you and your team get to the nitty-gritty work of creating the product. Make sure you set a clear deadline, which for a sprint is usually between two-to-four weeks. After this time, move on. Don't get stuck in this phase.

Crafting the product vision will give you perspective on the product, especially what it is and what value it brings to whom. Usually, you are building a product for a target audience. And often, you are not part of this target audience yourself. Therefore, don't just "think" of what the customer might like, ask them instead. The team has to be clear who the product is for, who the target audience is, and what to deliver based on the data gathered from the target audience. To illustrate this further, let's say you are leading a project for a home security company. The company wants to develop a new security system easily employed by the elderly.

Before conducting the whole project, some staff from the security company should have spoken with the target audience and gained solicited opinions, feedback, and insights. These can then be discussed with the Scrum team before it creates in the product vision. Remember, this is not the time to get into various details. The product vision is dynamic too, but everyone in the team should at least know the target audience and purpose of doing the project. Below I'll list some qualities for crafting an awesome product vision:

- **Make it broad and inspiring.** It is impossible to be specific in the beginning phases of any project. This is what makes the waterfall method so sloppy. People try to predict how everything in the project may unfold. This is simply impossible. There are always multiple variables that change. So, make the product vision broad but inspiring, so that people have a general idea of what they are trying to achieve and why. Make the "why" very clear and robust, and you will win.

- **Clear and stable.** Although employing various buzzwords seems to be a trend nowadays, don't fool yourself into thinking this should be the case for the product vision. Avoid complicated, unnecessarily long, and tedious language. Instead, be different, and make it clear and easily consumable for all team members, be they experienced or just starting in their role. Also, the specific vision should not be changed too often, if at all. It is better to keep it stable to avoid confusion among team members.

- **Short and sweet.** In the failing waterfall method, we see that professionals with high salaries make documents that are countless pages long. And what do you think the problem is with all of this? Well, let me tell you; the biggest problem is that no one reads all these documents. They are too long, complicated, and hidden from the crowd. Usually, after creating these documents, they end up somewhere in a drawer collecting dust. So, keep the product vision as short as possible. Why not do it on only one page that you can print out for others to see?

To keep these qualities fresh and the product vision in mind, you can discuss it during sprint reviews. Besides, I would like to share an example of how an inspiring product vision can come about. There was a CEO of a medical software and hardware company that helped craft the product vision in projects. Before crafting any plan, he would begin with a powerful speech about his son, who was very sick at birth. Due to his son's sickness, the CEO was often present at the hospital, where he was surprised to find many processes were very inefficient, ineffective, and slow. This fueled him to build his company to make the processes in hospitals quicker, more effective, and efficient to help more people in less time. This makes for a very inspiring message that pretty much any reasonable human being can relate to. Also, the product vision was placed on a single page and stuck on a wall near the Scrum team. Thus, it was always clear who the team was serving and how they were serving them.

The Initial Product Backlog

As explained earlier, the Product Backlog contains all the requirements of the product. Basically, it includes all items that you and your team should take care of to deliver the product and create value for the customer and organization. Its dynamic nature lends itself to the ever-changing markets, technology, and customer desires. The most valuable items are always on the top. These should be taken care of before anything else. The Product Owner has placed these items at the top for a reason, so get to work on these first and tackle them as quickly and efficiently as possible. The Product Backlog is prioritized by value, and it is more detailed on the top and less at the bottom. Eventually, you will get more detail when items move to the top. Although the Product Owner manages and prioritizes all backlog items, it's the Development Team that estimates the tasks, because they are the ones doing the work.

Generally, your Product Backlog contains user requirements, new features or enhancements and their descriptions, and technical requirements when things in the infrastructure need to change. These types of technical requirements aren't directly related to the customer. Make sure you limit them as much as you can. This doesn't mean that you should overlook them, not at all. Take them into account, but don't ever think they are the main reason why the work needs to be done. Always keep yourself and the team focused on the product vision. Without doing so, the team ends up forgetting this end goal due to the day-to-day work they are performing. Always find time to reflect and briefly look at the big picture. The same goes for bugs in the current system that block further development of the incremental product. You need to limit the bugs to remain focused on what the user or customer sees as value. I have said this multiple times already, but it's vital to reiterate: be certain that the Product Backlog is the *only* source of everything that needs to be done for the product. It includes everything from user and technical requirements, all the way to the various bugs that need to be fixed. So, there is no room for extra documents or places to jot down issues. This makes the whole process significantly more transparent. Transparency is important for any worthwhile project.

After filling the Product Backlog, the next step is prioritizing and estimating all backlog items. The key element in prioritizing anything is to make sure the item with the most value is at the top. In this context, with the value, we mean business value, namely the monetary aspect of things: increasing revenue. Another critical element is to group requirements wherever you see fit. Usually, various requirements or user stories are too big to tackle in a single sprint. Thus, you are better off splitting them. Less frequently, you might find similar or smaller requirements. Check if they can be bundled together and can be tackled in a sprint. Finally, I highly advise you to assign a business value metric to each story. Follow these steps carefully:

- Jot down all your user stories on Post-it Notes.
- Place the Post-it Notes on a wall or whiteboard. Just stick them right on the wall without thinking about any order.
- Now, you have ten "value" points to spend on each user story. Which user story gets the most points? When you can answer this question, place that user story on top, followed by the next user story with the highest score.

- If you are working in a larger organization or with a ton of user stories, ten "value" points may not be enough. Thus, you can take a broader range, such as 500 points or even 1000 points. The point is that you end up with a clearly prioritized Product Backlog based on the business value.

Tackling the most valuable item on the backlog might not be something that is needed in the initial stages of the software. In traditional project methodologies, you have no option but to develop all elements of the initial phase before moving on. Fortunately, with Scrum, you can develop the product increment with the most value first, be it something belonging to the initial phase of any other phase.

User Stories

User stories are necessary for gathering user requirements. This helps to form the requirements. A user story is short and simple; a watered-down requirement. Don't make them too detailed. Make sure they are made from the user's perspective. Usually, the developers are not the ones interacting with the product, so make sure the results or requirements for the users are formulated correctly. User stories make it easy to discuss the product, so focus on this. There is a general format to a user story, namely: As a <insert role> I want a <feature> so that <benefit>. Let's fill it in: "As an online clothes shopper I want to be able to search the website catalog so that I can find clothes to purchase." In this example, we used the role of "online clothes shopper," however, most websites are used by various people. So, for this example, you could also have a user who is a "discount clothes shopper." Based on the role, variables can be improved and innovated to match the specific user's desires. To do so properly, make a persona for each of the different types of users interacting with the website. A persona is a fictional representation of a specific user, such as "Bob TheDiscountShopper." Afterward, you can fill their persona by writing about their age, gender, motivations, goals, personality, and more characteristics that seem necessary.

Often, user stories are written on cards. You can put the user stories on the wall or whiteboard, but for larger organizations, they should be available digitally as well. Related to this are the "conditions of satisfaction," which have the following qualities: they are required for acceptance; represent tests; and are specific. Detailed items can be addressed later, but first, the specifics or foundations need to be in place. It doesn't change the significance of the effort when something can be addressed at a later time, such as the color of a button.

For instance, for the previously mentioned user story about the "online clothes shopper" searching the website, we can name various conditions of satisfaction. Think of things such as: match against item title; category; description; and keywords. Another requirement could be advanced search techniques like quotes. Furthermore, something like "Results should return in less than three seconds," is an example of a condition that you'll see in practice.

The conditions can be placed on the back of the user stories as steps on how to demo the user story or increment product. For example: "Open search page, enter various keywords, initiate search," et cetera. There are qualities of good user stories found in the acronym INVEST:

- **Independent.** When organizations are starting with the implementation of Scrum, they tend to craft user stories that depend on other user stories. Beware that you don't make this mistake and double-check that the user story can be fulfilled without the need for another user story.

- **Negotiable.** Dealing with user stories helps us deal with discussions. By discussing elements of the user stories, we can negotiate how to further the user story, if something is missing or if elements are superfluous.

- **Valuable.** Every user story should contain something of value for the customer, organization, or both. If this isn't the case, the user story is wrong and needs correction. If the user story contains no clear-cut benefit, it is difficult to prioritize it, compared to other stories.

- **Estimate.** The team gives an estimation of the amount of work and time needed per user story.

- **Small.** Make sure multiple stories can be completed in a single sprint. You do this by keeping the user stories small.

- **Testable.** The conditions of satisfaction aren't only documented and then put away. Nope, they're also tested. When the Scrum team has done an excellent job formulating the conditions of satisfaction linked to a user story, it is easy for testers to check that the resulting features are working exactly the way the user needs.

It can sometimes be difficult to manage ideas in an organization. Ideas seem to appear from all branches and all kinds of professionals. Usually, these ideas can be used to formulate user stories, but an idea on its own is not enough. Frequently, ideas can be combined, added to, or reduced in ways that are useful for the project. Dealing with all these ideas can be a hurdle, but Scrum makes it simpler. Scrum teaches us the hierarchy to better categorize them. The hierarchy is:

> • **Theme.** At the top of the hierarchy we find the theme. The theme is the overarching subject we focus on during an entire product release and sometimes even various releases. A theme is large and could never be put into multiple sprints, let alone a single sprint. Thus, it needs to be refined and reduced in size.

> • **Epic.** This is the name for user stories that don't fit into a single sprint. The epic user stories could be achieved in various sprints, but they need to be broken down to take care of finishing them in parts, one step (or sprint) at a time.

> • **User story.** Finally, we reach something that can be finished in a single sprint, the user story. Generally, in my experience, we always solved multiple user stories in a sprint, at least two. If this seems impossible for your team, make sure there is a way to split your user stories even further. Most of the time, it will be possible to do this.

Thus, before the Product Backlog is filled, make sure there is no way to reduce the size of the user story. Remember that it is better to have a smaller user story, easily achieved in a sprint, than a user story that could easily take up two, three, or even four sprints! During the sprints, always remember that developing goes iteratively and that you release the work incrementally. Sprint after sprint, we aim to develop or solve parts of a big puzzle. The process of developing parts of the puzzle is iterative. When a more substantial part of the puzzle is completed, such as the entire top section, we release this. Numerous sprints later, we release the middle section and bottom section. This process is incremental and is done until the puzzle is completed.

Crafting the Release Plan

Now that the Product Backlog is prioritized and estimated and the All-Star

Scrum team is ready to go, we are prepared to create the release plan. When crafting the release plan, think of the term "velocity." Velocity within Scrum is a measurement of the amount of work the Scrum team can finish each sprint. Put differently, it is all about the answer to the question: "How many of these backlog items can we do in one sprint?" Take a look at the image below.

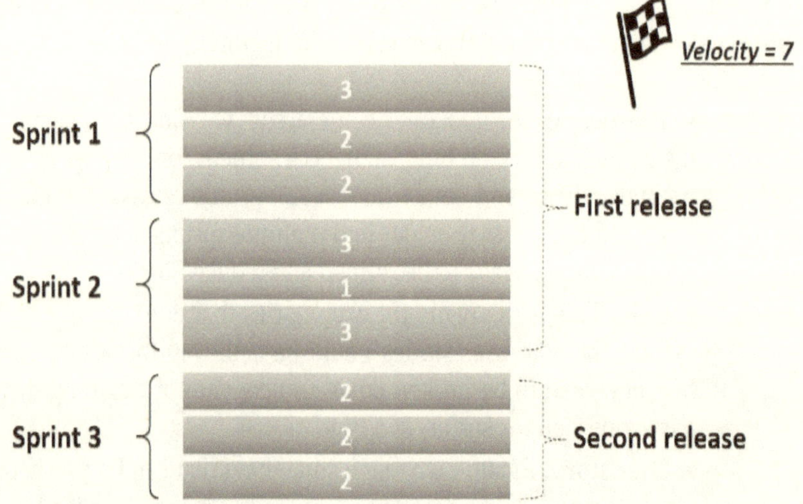

Don't worry if the image seems overwhelming. It isn't as complicated as it may look. Let me explain. In this image, the Scrum team is using a velocity of seven. This means that no more than seven story points can be taken up for each sprint. The velocity of seven is based on previous experience with the Scrum team. If your team is completely new to Scrum, start off with a lower velocity of five, to get a feeling of how the method works. In this example, the velocity is seven, and as you see, in sprint one the total amount of items that we take on is seven (3+2+2 = 7). The same goes for sprint two (3 + 1 + 3 = 7). However, in sprint three we see something different, the number is six (2 + 2 + 2 = 6). Why? Because if we take the "one" from sprint two and change it with the "two" from sprint three we would end up with "eight," thus exceeding the velocity. So, plan based on the experiences of previous projects. After the sprints are divided, the team can figure out when something can be released. In this example, two releases are marked: one at the end of sprint two and one at the end of sprint three. Thus, the team and stakeholders know when they can expect to release a product increment. Often, the Scrum team has a strict deadline before a release. This deadline might be set after sprint one, for instance. By discussing the current progress, stakeholders can point out if certain story points need to be in the next release. Then the team can adapt and move these story points to the top to take care of them earlier. But beware, make sure the velocity is not exceeded.

You might still be wondering how to go about figuring out your team's velocity when you are just starting with Scrum. Well, although it is not ideal to figure out your velocity without empirical data, it may be handy to calculate your initial velocity if there isn't any empirical data available. Here is the formula to calculate the velocity:

- Take the number of people in your Scrum team; let's say we have **six** professionals (P).
- Afterward, count the number of sprints and weeks of each sprint, let's say we have three sprints of **2** weeks each (S).
- Now, count the days the team is available to work on these sprints, for this example, **39** (D).
- Take a factor of ¼.
- Fill in the formula $P^2 + S^2 + D^2 \div \frac{1}{4} = V^2$

I hope you figured out that this is absolutely ludicrous and not at all possible! Although many people think figuring out these metrics is quickly done by filling out a formula, they are far from right. For a beginning Scrum team, there are immense amounts of variables you have to take into consideration. No formula on earth can fix that for you, except for trying, failing, repairing, and doing it again. So, instead of filling in this silly formula, have a discussion with your newly formed Scrum team about the backlog. Give each team member her/his say in terms of the various backlog items and how long they think the items will take. Then let all team members discuss the options and allow them to identify how many backlog items they can get done, for example: the top three items on the backlog. Then add up the number of story points of these three items, let's say the amount is nine. Then take that amount as your initial velocity for the first sprint. There is a high probability it is still wrong. But hey, we now at least have some empirical data we can learn from and do better next sprint. Right? During your Scrum project journey, more data will be gathered. After a couple of sprints, you can quickly figure out the velocity for best-case scenarios and for worst-case scenarios. When an item is very pressing and needs to get done, you can look at the velocity for the worst-case scenario (the worst the team has ever done), so that the team is confident they can get it done.

Chapter 8: Understanding Scrum Metrics

Scrum metrics are instruments for measuring the Scrum process. But what can you precisely measure? Well, you can measure a lot. Think about things like the development process; quality of work; productivity; predictability; and the products being developed. Scrum metrics focus on the value that is delivered to the customer. Besides Scrum metrics, there are also other agile metrics you should know about, namely:

- Kanban metrics: focus on the workflow, organizing and prioritizing work to complete this. A widely used metric is cumulative flow.
- Lean metrics: focus on valuation metrics from the entire organization to the customer and eliminating "waste." Popular metrics are lead time and cycle time.

But the metrics we address in this chapter are Scrum metrics, which focus on the predictability of a working product (or product increment) for customers. Commonly used metrics are team velocity and the burndown chart. We should care about these metrics, because by measuring you gain valuable insights into multiple Sprints, versions, when bugs are discovered, and more.

There are various elements that make metrics useful. Without these elements it is difficult to garner benefits from them. Consider the following:

- **Ensure that you use metrics that result in meaningful conversations.** For instance, after crafting the burndown chart, the team members can come together and discuss the current trajectory from time to time. Thus, they can figure out if meaningful progress is made.

- **The benefit of metrics is that you make the whole process more empirical.** Scrum relies on empirical data to a large degree. Thus, if you have metrics that support the gathering of this data, you support the whole Scrum process too. Think of data such as the speed at which a particular test can be performed.

- **Check if metrics can be combined.** Sometimes it is useful to combine various metrics to get a clearer image of the Development

Team's progress. Think of metrics like productivity (e.g., the number of tasks completed in a specific time) and velocity.

• **Ensure that the entire team comes up with the metrics.** The Development Team has a say concerning selecting the metrics. The Scrum Master and Product Owner also have a say. They can proffer Key Performance Indicators (KPI's) or metrics, like velocity.

• **Don't complicate things!** When you pick metrics, make them quickly understood. Also, calculating them shouldn't involve any advanced math.

Burndown and Burn-up Charts

You now might have the question, "How can I make a burndown chart?" To start measuring the team's progress. Well, that is pretty simple, by following these steps:

1. Draw a y-axis and an x-axis on a flip chart, whiteboard, or using any software.
2. Plot the story points or "work" on the y-axis. This is the sum of task-estimates in, for example, days.
3. Plot the time on the x-axis and include the total duration of your sprint or project. This is the iteration timeline, usually in days.
4. During your project, you measure from $t = 0$ at fixed intervals how much "work" has been delivered. In a sprint that takes fifteen days, you can measure at intervals of three days, for instance.
5. What remains is some basic math, namely working out the: Total – Delivered. Not very advanced for most people, I hope.
6. Now is the time to plot the result of this calculation in the graph. The points marked out for "Actual Tasks Remaining" should result in a downward line. The line shows per point how much "work" remains of the total you started with.

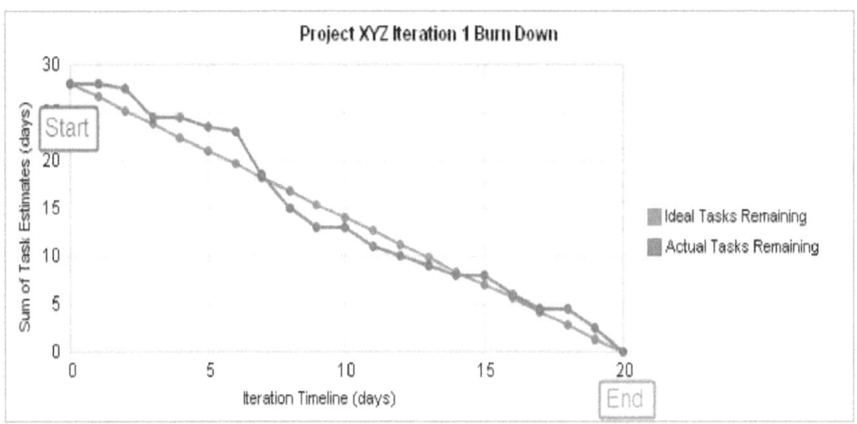

The descending "Actual Tasks Remaining" line follows the measurement points that you plot yourself in the graph. The "ideal" line in a burndown runs diagonally from the top left corner to the bottom right corner. The "ideal line" indicates precisely the average number of points that must be "delivered" per day to reach the project or sprint goal in the set time. When reading the graph, use the "ideal line" as follows:

- The distance between the real line (below or above) and the ideal line indicates how much the project is ahead or behind in the planning.
- If the actual line is above the ideal line, the project is running behind schedule.
- If the actual line is below the ideal line, the project is ahead of schedule.

Besides the burndown charts, you and your team can also decide to work with "burn-up" charts. The main difference between the two is that instead of tracking how many tasks are remaining, we track how many tasks or how much "work" we have finished. Doing so makes the curve go up instead of down. A burn-up chart comes in handy when the scope decreases or when the team happens to complete some work and shortly after finds that there is more work to get through for a new client, for instance. A burn-up chart makes these events more evident because progress is tracked independently based on how the scope of work changes or how the sum of task-estimates— or the "work" to be done—changes. In a burndown chart, there is no room to change the x-axis. With a burn-up chart, this is possible, because the "desired line" of ideal tasks remaining is a horizontal line that can change around. Take these steps to make a burn-up chart:

1. You plot the story points or "work" on the y-axis.
2. You plot the time on the x-axis and include the total duration of your sprint or project.
3. During your project, you plot the total amount of story points or "work" in the graph from $t = 0$ at fixed intervals.
4. At the same intervals, you measure how much "work" has been done, and you also plot this in the graph.
5. This then results in two lines: the relatively horizontal "total amount of work" line and the ascending "completed work" line. A burn-up chart clearly shows the completed "work" relating to the total amount of "work" in your project. The project is complete when the lines intersect.

This is how you make a burn-up chart. You can use any method you would like to make it clear, such as drawing it out on a whiteboard or creating one in Excel. Now that we have a burn-up chart, we should use and read it properly. The Development Team should compare the amount of work that is completed against the total amount of work in the project every working day. The distance between the two lines is, therefore, the amount of work that remains. As said, the project is completed when the two lines meet, i.e., intersect. The "ideal line" indicates precisely the average number of points, i.e., tasks that must be delivered each day to reach the project or sprint goal in the stipulated time. When reading the graph, use the "ideal line" as follows:

- The distance between the "completed work" line (below or above) and the ideal line indicates how much the project is ahead or behind on the schedule.
- When the line for "completed work" is below the ideal line, the project is running behind schedule.
- When the line for "completed work" is above the ideal line, the project is ahead of schedule.

Furthermore, the line that indicates the total amount of work makes evident when work has been added to or removed from the sprint. If "work" is added to the project or sprint, the distance between the two lines becomes greater.

The influence on the planned end date is, therefore, clear. To make this clear, you draw a trend line that is based on the average amount of points that are delivered by the team. The new end date of your project is where this trend line crosses the (increased) total work line.

You can experiment with both burndown charts and burn-up charts. Mainly if the Development Team includes members who are new to Scrum, this is important. Besides experimenting with both methods to see what fits well with the team, some projects demand a particular approach. There are projects which revolve more around one of these two questions:

> 1. Do you want to monitor the progress of your sprint in detail and share it with customers and stakeholders regularly?
>
> 2. Do you want to make progress visible in the simplest way for the team and those directly involved?

A burn-up chart shows the completed work and the total project size in one overview. In contrast to the burndown chart that only shows the remaining work with one line. Because the total project size with the burn-up chart is clear, you get answers to questions such as: has too much new "work" been added?

This information is useful for identifying and resolving problems within the sprint, e.g., a customer who is constantly changing the scope, putting the end-result at risk. Use the burn-up chart for reasons such as:

- **You want to show regular updates on how the sprint is progressing.** If you regularly present the project progress to the same audience, for example, weekly progress meetings for customers, use the burn-up chart. This makes it easy to see the progress, including the additional work and the consequences it brings forth.
- **If the scope has a "dynamic" character.** When "work" is suddenly added to the sprint, a burn-up chart is more valuable. In some projects, the customer can be very pressing and ask for additional functionality. Besides, some tasks could be removed if they are not necessary anymore. Usually, the latter is caused by unexpected problems with costs, budget, and time. In a project where customers add or remove a lot of work during the project, a burndown chart is not a complete representation of the progress.
- **When the scope expands.** When the project scope is gradually

expanded, sometimes far beyond the original framework, a burn-up chart is more useful. With a burndown chart, it would appear as if little progress has been made by the team, which is not necessarily the case, i.e., the scope could have been expanded. Thus, a burn-up makes the issue at hand much clearer for the customer. This makes the issue discussable and allows for a quick adjustment whenever necessary.

Why should you use a burndown chart instead? There are a few reasons:

- **Very easy to employ.** Burndown charts are easy to make and follow up. They have an obvious line to show when the project or sprint is finished. Starting with a burndown chart is particularly advisable for new Development Teams. Why? Because, every team member will understand the graph with little to no explanation. Thus, making a burndown chart for presentations is desirable for customers or stakeholders from a non-technical background. However, don't forget that the burndown chart does leave out some crucial information. It doesn't tell the whole story of the sprint. So, for more complex sprints, with changes in the scope, this isn't very useful.
- **The scope has a "static" character.** Burndown charts are usually employed with projects where the project doesn't change at all. Projects or sprints with a fixed scope don't need the additional information of a burn-up chart. This would make the project unnecessarily more complex. Maybe it wouldn't even be more complex, but just not useful at all. When things are simple, keep them simple. Any unneeded variable you add to your plate makes the project exponentially more difficult.
- **Helps keep the momentum high.** The burndown chart can keep the team's momentum going and can motivate them to push through until the end of the sprint. When people are starting with Scrum, these quick wins build confidence. And confidence is the quality that the Development Team needs to deliver quality product increments. When the team members are all more confident, they create, provide, and bring more value to the customer.

Chapter 9: How to Excel and Common Mistakes

There are various ways to excel as a Scrum Master or Product Owner. Multiple ways have been addressed already, such as being transparent with your team members and being open to their opinions and feedback. In Scrum, you always strive to make the users or customers happy. Preferably with something tangible. Thus, when you are new to Scrum, don't take on a large and complex development job from the get-go. To excel, take a business project whereby the users don't care about all the intricacies of the product. For instance, if you are creating a marketing tool for the Marketing Department, they are less interested in how the programming code ties together and how well it is documented. They just want a working tool, that's it. Also, starting with a small project gives you more confidence. Ensure success first and scale up afterward. Start with one product, one Product Owner, one Scrum Master, and a team of five or six people. Don't think that you can convert the entire company in one day.

So, a short project is good to start with, but by short, I don't mean *very* short. A very short project probably gives you too little time to see Scrum work adequately. Then the success of the project depends more on some "heroic deeds" by some team members. Moreover, small projects are not sufficiently representative. If the project is a success, then you will still be told: "We would get the same results with the old method." I would say a project of about two to three months is fine to start using Scrum with.

A common mistake we notice is that new Scrum teams take on projects that aren't "real." In other words, a project whereby it doesn't matter if things go wrong. That is pretty much the worst thing you can do. An unimportant project receives no attention, and therefore no focus, no stakeholder involvement, and no worthwhile meetings. With an important project, it is much easier to get people motivated. And you don't have to be afraid that things will go wrong: as a team member you already have skills and are able to achieve goals, no matter which method you use. Always add Scrum elements that make you better, never make a mistake to think that Scrum is a goal in itself. It's not. Instead, think of it as a way to work smarter, not harder. The business side of your organization is sensitive to this.

Another mistake a lot of Scrum Masters and Product Owners make is that they don't garner support from all layers of the organization. If everyone is new to the Scrum scene, it is possible to hire an experienced Scrum Master to start things off and to make sure that everyone supports the Scrum initiative. It is an experiment worth trying. Make sure that the Product Owner and the team members have a good reason to make it a personal success.

Also, don't be the one who is afraid of failing. You can never excel if you are allergic to failing. Instead, embrace the fact that failing is part of the process to gain the mastery you wish for. Then take the time to become better at Scrum and adhere to the rules of the game. Especially with a first Scrum adventure, there is a tendency to exaggerate. Think that every team goes through phases to become magnificent. This ties together with communication and transparency. Make sure you make all the problems visible. Scrum itself is not going to solve your problems, but it makes them readily available and puts them "right in your face." By gaining this visibility, you have the best chance that you can to mobilize people to solve the impediments.

Don't make the mistake of thinking that change happens overnight. I always show some character and a strong personality when needed. It is better to ask for forgiveness afterward than to ask for permission in advance. No room to discuss with stakeholders? Well, can you go outside and discuss it there? Are no users or customers visiting the demo? Find them and invite them for some coffee, and show what has been achieved. When you achieve anything meaningful during the implementation of Scrum, celebrate it! Give presentations, organize a dinner, or give a small gift to the team members and stakeholders. This will fire them up and help them in their growth.

Finally, another common mistake is completely forgetting the use of technology to facilitate the Scrum process. Nowadays, a myriad of Scrum tools exist which can set you up for success. Why not make use of these to keep the processes more streamlined? In the next chapter, various tools will be addressed that can fast-track your project's desired outcome.

Chapter 10: Scrum Software Tools for Workflow Management

In recent years, an enormous amount of software tools have found a place in the emerging technology market to ease the Scrum process. For a team that is not "co-located" and/or is not working full-time with Scrum, the benefits of a digital solution are crystal clear. Handy integrations with other digital platforms also add extra functionality and value. A physical project board still works best in terms of interaction, for example, during the stand-up meetings. And when creating transparency and insight into project status, a whiteboard filled with colorful Post-its is hard to beat.

From a functional point of view, the range of digital Scrum tools can be subdivided into: collaboration & documentation tools, such as Portals, where "knowledge" can be stored and shared; workflow managers, which are useful for managing workflows by using digital project boards; and messaging tools, to enable and streamline team communication. Also, it is possible to find tools that are a combination of these categories.

Some software companies specialize in one aspect, and others combine functionality. The advantage of an integrated solution is that everything "seamlessly" intertwines; a disadvantage of an integrated solution is that it sometimes contains a component that doesn't function well in the scheme of things. In the latter case, a combination of specific digital Scrum tools might have been a better solution. The prominent providers of integrated solutions compete with each other on functionality and price. Most tiers and price models are not far apart. The digital options often start with a free version with stripped-down functionality and a limited number of users. The first subsequent "Tier" (price level) will then be around $10 per user from fifteen users. From a functional point of view, the range of digital Scrum and agile tools is so extensive that it is difficult for the starting "Agilist" to make a choice, let alone the right choice. That is why we have given an overview here. This is, of course, not an exhaustive overview; the range of literally hundreds of applications is too broad for that.

We start with some entry-level tools that are completely free: Trello and Slack. If ease of use and progress reporting play an essential role from the start, Asana is a good choice for smaller organizations and teams. A more robust solution is advisable for larger organizations that want to implement Agile broadly. The Atlassian combination of Jira Core, Confluence, and

Stride, offers a solution here. Take into account the fact that switching from your current application to a new application can be a great hassle. So, always make a very thoughtful decision and communicate things with the Scrum team. Here are the options to get yourself started:

- **Option 1: Use the free and simple tools Trello and Slack for small teams.** Trello is a workflow manager based on the Kanban board and has a free version with unlimited boards, lists, maps, members, checklists, and attachments. Trello is very easy to use and hardly has a learning curve. The instructional material is well put together and quickly provides sufficient knowledge to be able to use the platform. The free version is highly recommended for starting Scrum teams. Slack is a Messaging tool and has a free version with an unlimited duration, but limited storage of 5GB and 10,000 messages. However, this is more than enough for small teams who want to try it out. The Slack Standard Plan costs around $7.25 per user per month.

- **Option 2: The Scrum tool Asana is great for larger teams, and when an organization wants to conduct multiple projects simultaneously.** Asana is a widely used Scrum tool that combines a workflow and project manager with excellent team messaging and collaboration functionality. Asana has good reporting functions that provide insight into the progress of the various projects with a single click of the mouse. There is a free plan for up to fifteen users and a premium plan for around $9.99 per user per month.

- **Option 3: Scaling larger, more complex projects with enterprise-quality software by Atlassian.** This comes with a combination of the following Atlassian products: Stride (a very comprehensive messaging tool); Confluence (a documentation and collaboration portal); and Jira Core (a workflow and project management application).

The team could also use things like G Suite by Google. Using G Suite by Google can help create Scrum artifacts and keep only one version of them. Thus, these documents can be updated in real-time, and team members can be invited to have read-only rights. This is important because the Product Backlog and other artifacts are very dynamic. During a project, things can change. Thus, requirements or backlog items in the Product Backlog can change too, and it might be useful for the team to have digital access to this information.

www.ingramcontent.com/pod-product-compliance
Lightning Source LLC
Chambersburg PA
CBHW020500160726
47991CB00007B/2746